Step-By-Step To Organic Vegetable Growing

Significant portions of this book, together with many of the illustrations, first appeared in the book "THE NEW ENGLAND VEGETABLE GARDEN," published by the Countryman Press in 1957.

Step-by-Step to
ORGANIC VEGETABLE GROWING

By
SAMUEL R. OGDEN

RODALE PRESS, Inc. Emmaus, Pa.

Standard Book Number 0-87857-001-2
Library of Congress Catalog Card Number 74-149930
Copyright 1971 by Samuel Ogden
All rights reserved
Printed in the United States
NINTH PRINTING – NOVEMBER 1975

OB-579

Contents

To the Campbell Girls
my wife "Mamie" who kept me at work on the book,
and "Miss Fannie," her sister, who kept
herself at work in the garden. Without
the one, no book; without the
other, no garden.

Preface

THIS PRESENT VOLUME is the reincarnation of my *New England Vegetable Garden* which was published in 1957 by the Countryman Press of Woodstock, Vermont. That book had a brief and unhappy career, and soon was out of print, whereupon, in ever increasing numbers, as time went on, those who had seen the book, either in the libraries or in their friends' hands, wrote in to find out where they could get a copy for their own use. For reasons not quite clear to me, which have no real bearing on this current enterprise, it became impossible to obtain a copy of the book. When, finally a dealer in rare books wrote to me asking where some copies could be obtained, I decided that something should be done, and the present happy collaboration with Rodale Press is the outcome of that resolve.

In the first place I should like to make it clear that while this work is a handbook for the growing of vegetables without the use of chemicals of any sort, it makes no pretense to be comprehensive as far as species and varieties are concerned. I re-solved (at the time of writing the book in the first place) that I would not speak of any matters of which I had not first hand information, the results of my own actual experience in preparing the soil, in planting, in cultivating and in descriptions of the vegetables grown. Since the climate of this high valley condemns us to an extraordinarily short season, many species can not be grown and many varieties have to be selected because of the speed with which they can be brought to maturity. But let me assure you that the one who picks up this book hopefully, need not turn away from it because lima beans or pumpkins or country gentleman sweet corn are not listed among the vegetables grown.

The most important ingredient in a first-prize vegetable garden is the soil (and in the making of this, there is not one set of rules for one vegetable and another for a different kind). The methods for making the soil into something in which all plants thrive and proliferate are universal, without the need for modification with respect to location or climate. The three basic conditions which must be met if best results are to be obtained are: the perfection of the soil, the proper preparation of the seed bed, and the cultivation of the growing plants. The work you are now looking at is principally a handbook for covering these procedures and techniques; a treatise, so to speak, for learning how to play any tune you choose to on the instrument which is your kitchen garden.

The *Encyclopedia of Organic Gardening* likewise published by Rodale Press will supply answers to questions which arise outside of the scope of this work. Furthermore, in most instances the package in which the grower places his seeds includes instructions for the proper culture of its contents.

In the fourteen intervening years, I have discovered no inconsistencies or errors in the original work. Thus the present volume, while modified here and there (in inconsequential matters) remains substantially the same as originally published, with the exception that many of the photographs appearing herewith are fresh and new. As these will testify, the methods I advocate do produce splendid results, and in all the years intervening, there has never been a garden season which has fallen below the standards pictured here.

The extraordinary change in the public point of view concerning the use of poisonous pesticides and herbicides, and the conviction that

serious harm results from the run-off from soils treated with soluble phosphates and nitrates, which has come about since Rachel Carson wrote *Silent Spring* in 1962 have resulted in the rejection of philosophies and practices which, up until now, have been calmly taken very much for granted. The realization of the dangers to ourselves and our environment inherent in much of man's treatment of nature, has induced a tremendous upsurge of interest in the kinds of food we eat, and in the manner in which they are grown. Matters which not so long ago were brushed aside as the vaporings of fanatics, are now being considered seriously by more and more people, and as a result there is a large demand now for organically grown vegetables.

One aspect of this tremendous interest is that more and more young people are attempting to return to the soil as a way of life. The deteriorating environment of the suburbs and the cities is forcing many to take a second look at what we humans are doing to ourselves, and an ever increasing number of young people are saying "to hell with it," and are taking to the country. This is a subject of great interest but one which can not be gone into here. I mention it, nevertheless, because it has had a very noticeable impact on the pattern of life here in southern Vermont.

On numerous occasions I have received delegations of young people, oddly garbed and haired in many instances, but none lacking in cleanliness and good manners, who have sought me out to visit the garden, to ask where they can get a copy of the garden book, and finally to find out from me how they might go about achieving their hopes and desires. So, it is to these people, and to all others who want to cultivate their "pieces of land not so very large" that I turn in hope that they may find help herein.

S.R.O.

Landgrove, Vermont

March, 1971

CHAPTER ONE

Good Reasons for a Garden

MANY YEARS AGO Horace wrote: "This is what I prayed for, a piece of land not so very large, where there is a garden, and near the house an ever-flowing spring of water, and above this a bit of woodland." The love of the soil, an urge to make it bear fruit, is a part of the heritage of most of us, whether we be city bred or transplanted thence, or dwellers in the country. Events which have taken place in the memories of us who are still living have changed the face of the earth and the lives and habits and thinking of most human beings, light-complected or dark, East or West, no matter who we may be or from what origins we may have sprung. In all of these changes the powerful sweep of the trend has been away from Horace's piece of land where there is a garden.

One of the most profound and far-reaching changes perhaps has been the change in attitude toward nature on the part of western civilizations. Now it is man's self-assumed destiny to master nature, to bend her powerful forces to his will, rather than as a part of nature, work with understanding of her unchangeable laws. As Faustus made his impassioned demand on magic, so we make ours on science, and with Faustus say: "I charge thee to wait upon me whilst I live, to do whatever Faustus shall command, be it to make the moon drop from her sphere, or the ocean to overwhelm the world." It seems to me that this lust for power is the evil of our age, and I believe that work with the earth, even on ever so small a scale, will help us cure this evil. I am not proposing that we should trade one brand of magic for another, for magic will not work. Nor will all who work with the earth in order that it may bear fruit find any cure therein. In fact, one who has written on growing a garden says: "I wait impatiently the lowdown on raising vegetables without benefit of soil or sunlight, in chemical solutions, and on insect control by light and sound." To one who has the disease to this degree, the garden plot is merely a proving ground for the miracles of science, an opportunity

to slam one more door in nature's face, and a place where he will have perhaps lost the last chance to gain a bit of wisdom and serenity.

No, to have a garden is no sure cure for any disease whether it be of the body or the spirit, but it can be a source of personal satisfaction and profit, and it may be solace against the distractions of our age as well. As one works with one's hands, the mind is left free for thought and meditation, and the surroundings of garden work are such that these contemplations tend in directions which at least seem to be profound and satisfying. There are so many needs for us to think nowadays, and there seems to be so little interest in or talent for thinking in our own lives or in the lives of those about us, that any occupation which gives us pause to think has a value far beyond its intrinsic worth. So, at the risk of being called out of order, inasmuch as this purports to be a book on vegetable gardening, it seems worth while to me to develop briefly my theory that to work with one's hands in the soil, to change the fallow to the fruitful, has, in this age of fears and instability, a value far beyond that of the material produced. Those who are willing to agree with me, or who want to get along with the business of gardening, may skip this and go along to future chapters where I will speak of the actual business of making a garden.

So many people are concerned nowadays with the breakdown of our morality and the threat to our physical world implicit in the hydrogen bomb, that the existence of a worsening condition in man's affairs is taken for granted by nearly everyone who studies the situation. In fact, one profound book written not long ago comes to my mind wherein the opening sentence is: "This is another book about the dissolution of the West." On the other hand, many who to themselves may admit to a feeling of apprehension, will still insist that all is well with the world; that, in fact, the general condition of mankind is much better now than at any point in history. Obviously both points of view cannot be true. Either the first group, and it is very much in the minority, is composed of a bunch of calamity howlers, or the rest, the most of us, have, like ostriches, stuck our heads in the sand. To be convinced of the perfectability of man and the reality of progress one merely has to float along with the materialistic and scientific pronouncements which are printed in our periodicals, taught in our schools and universities, and mouthed by our politicians. In so doing one does not really have to think, for are not all the great thinkers agreed? For another thing, and most important, the latter belief is much the pleasantest and easiest to accept.

Richard M. Weaver, the opening sentence of whose book *Ideas Have Consequences* I have just quoted above, finds the main difficulty of such discussions is in getting certain initial facts admitted. He says: "This difficulty is due in part to the prevailing Whig theory of history, with its belief that the most advanced point in time represents the point of highest development, aided no doubt by theories of evolution which suggest to the uncritical a kind of necessary passage from simple to complex. Yet the real trouble is found to lie deeper than this. It is the appalling problem, when one comes to actual cases, of getting men to distinguish between better and worse." It is my contention and conviction that the slow tempo of garden work and the meditation which of necessity goes along with it develops in the practitioner a sense of basic values, and these values must be established before it is possible to distinguish between better and worse. How many of us in our lives have not met persons whose serenity and wisdom were acknowledged by all, and yet whose educational and cultural backgrounds were of the briefest sort? This, then, is important. If gardening will by virtue of its practice tend to develop in us an understanding of nature and a perception of true worth, he who thus practices it may become a better and wiser person.

There are other things about a garden besides its produce that make gardening very much worth while. Of these I wrote in an earlier book substantially as follows:

In the time of our grandfathers and great-grandfathers, far more people, proportionately, lived in the country. In each family, from half-grown children to grandparents, everyone was busy from sunrise to sunset. They were busy living. They had to be sheltered from the weather, be clothed, fed, and to have some surplus of material goods laid up against the future.

They felled their trees and cut the timbers to frame the house or barn, hewing them into shape; they split shingles from spruce or hemlock blocks. They raised sheep, and clipped, cleaned, and carded the wool; they spun the wool into yarn and wove the yarn into the cloth from which they made their clothes. They grew flax and made their own linens. They made their soap with fat and wood ashes. They tapped the sugar maples in the spring and made their sugar. They battled with the soil, the rain and frosts and wind to raise wheat, which they ground into flour. From the flour they made their own bread. They raised flocks and herds that they might have wool and hides and meat. They made their own butter and cheese.

The land produced the things whereby they lived; there was a direct relationship between human life and the soil. Even in the cities this relationship was no doubt apparent and real. Now, in our times, in these days of concentration in the great cities, of specialization and industrialization, this feeling of closeness to the land has been completely lost. Yet it is an integral part of the race heritage of each one of us. For us to lose contact with the soil results in real unease and maladjustment; to recapture it affords profound joy and inward comfort.

These two things are vital: to have a chance to think and perhaps gain wisdom, and to achieve a renewal and readjustment through direct contact with nature. The way to these ends may be found, I submit, in this "piece of land not so very large, where there is a garden."

Besides these, there are the tangible results, the production of useful consumable goods which take the pressure off our budget, which we can feel and count and taste and measure. To the Ogden family this tangible result amounts to between thirteen and fifteen hundred dollars annually. This is higher than most, for while our garden is not in the proper sense a commercial garden, it is more than a simple family kitchen garden. We live in a small community the ranks of which are swelled every summer by vacationing families. Most of these have not the time, nor are they here at the proper time of year to garden themselves, so they are delighted at the opportunity to get fresh home-grown vegetables from us. But for anyone, even for the smallest family, the results are amazingly profitable. Records of the United States Department of Agriculture show that, under favorable conditions, the time spent in the garden yields a return equivalent to that obtained from a corresponding period devoted to regular employment. It has been estimated that a vegetable garden can be made to reduce the amount of money spent for food to the extent of five or ten per cent of the average salary.

In Vermont, and many other states, there is a slack time in the farmer's year toward the end of March and the beginning of April. Snow is still in the woods and frost is in the ground. The sun is hot and sap begins to run, but the frost and the snow and the mud will not permit the soil to be worked. So it is at this time the farmer taps his sugar maples and turns to the making of maple sugar and syrup. Often by dint of hard work — when the sap is running freely, he is at the sugarhouse boiling down the sap all night long — he turns out a product that has real cash value.

It requires from thirty to forty gallons of sap to produce one gallon of twelve-pound syrup, depending on conditions of season and weather. Under favorable circumstances, one tree will yield seven or eight gallons of sap, or about one quart of syrup. The price received for the syrup varies with the grade, season, and general market conditions. It is interesting to note that in 1942, when I wrote *How to Grow Food for Your Family*, I quoted the price as ranging from $1.25 to $2.50 per gallon. Last season the price ranged from $7.00 to $10.00 per gallon. So perhaps now the price received is more nearly a measure of the cost

of the article in terms of labor, but in any event the producer is not overpaid. The point is that the time expended on this very pleasant occupation is spent when there would otherwise be nothing productive to occupy the farmer's time. As it is, he receives hard cash at a time of the year when it is most welcome, and he has fun earning it.

The case of the home gardener might be considered a parallel one. Time wasted, or used on non-profitable hobbies, can well be spent in the garden, adding a substantial sum to the family income and at the same time fulfilling, in a most healthful and satisfying manner, the need that all of us have for a hobby.

In addition to all these things, the crowning advantage in having a family vegetable garden is the privilege of choosing the varieties that we prefer. The vegetables on sale at the greengrocer are produced commercially, fed with chemical fertilizers, and handled with machinery. Of necessity, the varieties have not been selected for their flavor or succulence, but for their ability to stand the rough treatment of machine methods of cultivation and harvesting, as well as their hardiness to endure long shipments and many handlings. The home gardener, on the other hand, has a wide choice alluringly displayed in the seed catalogs. He can choose on a purely personal basis, his selection governed only by the requirements of the family, and the exigencies of the climate in which he lives. Thus he can place all the emphasis on the quality and texture and flavor.

The problem of selecting the vegetables to grow, and the species to choose is confusing to the beginner. The start is made with professional advice, which is one of the purposes of this book, but time and experience, together with swapping results with other gardeners, will bring each, in the end, to his carefully selected list, a list which will be modified from year to year, for the true gardener will keep on experimenting to the end of his time.

A decidedly important reason for the home garden is that the vegetables can be gathered for immediate table use when they have reached precisely the proper stage of maturity to be at their best. If a customer requests a dozen ears of corn, I always ask at what meal the corn is to be served. If it is ordered in the morning and is not to be served until evening, I will not pick the corn until late in the afternoon. Peas and beans are best when they are young, and the pods are not filled out to their maximum. Here is a loss in volume and return which the commercial gardener cannot afford to suffer, but the home gardener can, and the difference in eating pleasure has to be experienced to be believed.

A garden grown as one should be, without chemical sprays or fertilizers, will produce vegetables which are superior in taste and quality. These attributes take color from the soil in which the vegetables are grown to an extent that is little recognized. This is particularly true

of leafy crops, the growth of which has been stimulated by the use of soluble nitrates. There is a health aspect resulting from the use of soluble chemicals which is receiving more and more attention from the agricultural experiment stations, but the pressures of commerce are so great in every direction that it would be naïve to expect that the results of these discoveries will modify in any drastic way the quality of the vegetables that come to our tables by way of the greengrocer. More of this later; in the meanwhile be assured that the only way to get the best in taste and health is to grow your vegetables yourself, unless you have a neighbor who will do it for you.

So then, let us look for that "piece of land not so very large" where we can make our garden.

CHAPTER TWO

Soil for the Garden

CHAUCER WROTE: "For out of the old fields as men saith, cometh all this new corn, from year to year," and these old fields, whether they be a patch of hard, compacted back yard, a strip of greensward, an ancient and weed-choked kitchen garden, or veritably an "old field" of some abandoned farm, are the starting point of our gardening enterprise, wherever our garden is to be grown. So we must first concern ourselves with the soil, its origins, its quality, and its texture. Plants can be grown without soil in aqueous solutions, but the devices of hydroponics are not for the gardener.

For our garden we must have soil and sunlight, which are the very foundations of life. Down through the ages the earth has been referred to as "mother earth," and with reason, for it is from the soil that all vegetation springs, and all life on the surface of the earth is nourished by this vegetation.

Soil is the loose material which covers the surface of the earth and is primarily the result of the disintegration of the rocks which make up the better part of the earth's crust. This disintegration has been caused by the action of heat and cold and water upon the rocks, by living organisms, and by the movement of wind and water. The whole process is too complicated to be discussed in detail, having taken place through aeons of time and involving many complex actions and reactions.

With the hope that I might be able to come up with some significant and helpful general relationship between the parent material from which the soil was derived and the nature of the soil itself, I spent many pleasant hours with *Soils and Men*, the 1938 Year Book of the United States Department of Agriculture. I found much of specialized interest, but little that had broad enough application so that one could say: "The nature of the bedrock here on this site is such that the derived soil will have this, that, or the other thing characteristic of it." In fact, in the chapter on "The Formation of Soils" I found this statement: "Many soils may be examined to the depth of two or three feet without finding any inkling as to the nature of the parent rock from which the soil material was derived." From this and other bits gleaned from the writings of the soil specialists, I came to the conclusion that there is very little, if any, relationship between the nature of the soil and the kind of parent rock which lies beneath the surface, and I returned to my previously held conviction, which came as the result of practical experience, that the most important aspect of a soil is its physical or mechanical condition. In other words, the important factors are not the derivation of the soil, or its classification, but rather its tex-

ture, its depth, its content of air and moisture, the presence or lack thereof of rocks and stones, etc.

To illustrate the point, let me use the example of my own garden, which is a very highly productive one. It is located on Podzol soil, which by identification is apt to be very poor, and is classified as "Stony Berkshire," which is just about the worst soil classification that there is in the United States. The underlying rock is Green Mountain Gneiss; in general, the soil material has been transported by the glacier and abounds in stones ranging in size from bungalow-sized boulders to small pebbles. Part of the garden has a sandy subsoil, and part of it is

ALLUVIAL SOIL

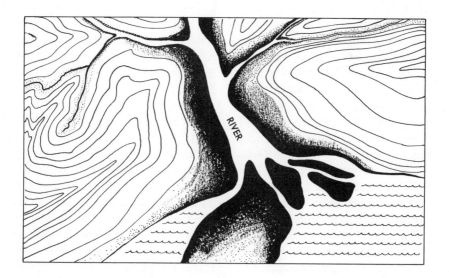

Typical mountain and valley system, showing how topsoil washed from the uplands creates deep sediment deposits characteristic of fertile "bottom" lands.

[9]

over impervious hardpan. But it is a good garden, and the fact that it varies in character due to the difference in subsoil gives me a chance to manipulate my crops accordingly.

An old professional, W. W. Rawson who published *Success in Market Gardening* in 1887, in speaking of soils does not mention soil classification or parent materials, but confines his observations to the physical and mechanical aspects of the soil. He says: "Rocky ground is of course and by all means to be avoided for garden crops, in view of the deep and uniform cultivation that it needs to receive. And low lands which required under-draining are adapted only to certain special crops, and involve heavy outlays to make them capable of profitable culture. Preferably to either, a sandy loam with a sandy or gravelly subsoil should be selected. Such land is far better than soil resting on clay, not only because its nature is warmer, but because it is naturally well drained." Note that these words of advice are directed to the commercial market gardener, not to the amateur. For the small kitchen garden, Rawson's generalities can be modified as we shall see, but the basic principles are sound.

There are, however, two geologic processes which have a direct bearing on the physical nature of the soil, and these in general are the action of ice and of water. Streams and rivers pick up soil materials in their courses from the mountains to the sea, and in this same great movement the solids are redeposited, thus forming the alluvial soils. In general, alluvial soils are highly productive, free from rocks and stones. In large river basins, the wide range of rocks and soils in the drainage area makes it likely that in these soils there will be an abundance of plant foods as well. A large proportion of the world's population depends on these rich alluvial soils for its foods.

More or less in contrast to the alluvial soils are the glacial soils. During the ice age much of the now temperate zone was covered deep with ice, and glaciers forming on the edge of the icecap flowed to the south, gouging and scouring the rocks, deepening the valleys, rounding the hills, and picking up an ever-increasing load of commingled clay, sand, gravel, and boulders as they moved along. With the shrinking of the icecap and the retreat of the glaciers, this accumulated cargo, known as glacial drift, was deposited on the face of the land, thus forming the glacial soils. These glacial soils, which sometimes are rich in plant foods, are likely to be sprinkled with large rocks and stones, and in many instances are gravelly or sandy. Those of us who live in northeastern United States are apt to find that our garden plots are located on glacial soil, for the glacier covered all of New England and New York, the northern part of New Jersey and Pennsylvania, most of Ohio, Indiana, and Illinois, and all of Michigan and Wisconsin. It is interesting to note

FURTHEST ADVANCE IN THE UNITED STATES OF THE LAST GREAT GLACIER (ABOUT 20,000 YEARS AGO)

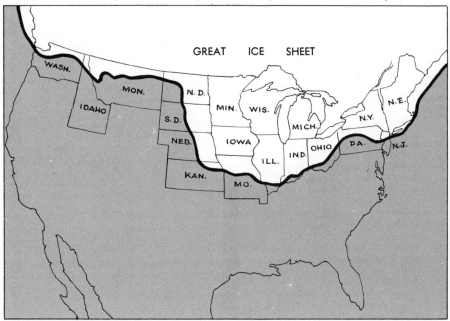

that the entire British Isles with the exception of the extreme southern tip of England were likewise covered by the glacier.

After the soil has been formed as primary or mineral soil, further action must take place before it is ready for our use. Further weathering takes place, primitive forms of vegetation appear, followed by more complicated forms. It is this vegetation which really transforms the material covering the earth into that which we know as soil. The decay of the vegetation, resulting in the building up of organic materials, the action of burrowing animals and insects, the appearance of bacteria and fungi, working together in com-

plicated harmony, finally bring the soil to the point where it is rich and friable and ready for our use. The making of the soil by nature has taken untold centuries; the marring of it by man can take place within the span of a single generation. It is the work of but a few years either to deplete the soil in fields or in gardens, or to make it deeper and richer and more productive.

When we choose the site for our garden we more or less have to take the soil as we find it, whether it be glacial or alluvial, podzolized or laterized, whether it be Stony Berkshire or Hagerstown Frederick; in all but exceptional cases it will not be perfect, either in plant food con-

THE BUILDING OF GLACIAL SOIL

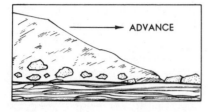 

Ice sheet advances, scours up rocky debris which becomes embedded in glacier.

As glacier retreats, debris which has been abraded in transit, is redeposited, called *glacial drift*.

Boulders begin to fracture and disintegrate from physical stress, expanding with heat, contracting with frost. Scoured by wind and eroded by water, boulders become pebbles, pebbles become sand or clay.

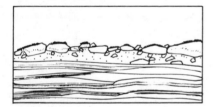 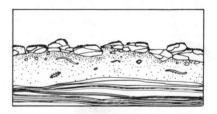

Primitive vegetation such as mosses, molds, lichens, introduce first organic deposits.

Bacteria and insects feeding on decay of vegetation . . . themselves die and decay.

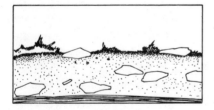 

More active vegetation takes hold, accelerates chemical cycle operated by plants, animals, weather. "Humus" impregnates original mineral structure.

Advanced vegetation and animal life deepen organic content of "parent" material. After centuries, so-called topsoil achieves "mature profile."

tent or physical structure. We must take that which is available and make it over into what we want it to be. The good gardener can take the most unprepossessing bit of land and change it from fallow waste into a lush and productive garden.

Starting out, two intertwined and continuing problems confront us. These tasks are to bring the soil into suitable physical condition, and to supply the plant foods necessary for vigorous growth; unless the soil is so shallow on top of base rock that it will not hold moisture, or soggy wet and so situated that it cannot be drained, these ends can be achieved.

Assuming that there is sufficient depth of soil and that, if wet, it can be drained, the initial job is to clear the land. If the soil is glacial, the presence of rocks and stones may present the first problem. In spite of the statement of W. W. Rawson, previously quoted, rocky or stony ground need not discourage us, for the limited area of ground necessary for a family garden, and the high productivity will justify spending time on rocky soil which the commercial gardener could not afford to give. Depending on tools available, large rocks can be removed. By large rocks I mean those larger than a man can handle and which require the use of a stump puller, a winch, or chain fall. Larger rocks which would take dynamite or the use of a bulldozer to move off had better be left in place, but to get rid of as many of the smaller ones as

humanly possible will be all to the good. A wall one hundred feet long and three feet high was built along the lower edge of my garden using rocks taken from the garden over a period of time, many of them larger than any man could handle alone.

Smaller rocks and stones can be carried or thrown off the garden as the plot is worked; as the largest of these are turned up by the spade or the plow, carry them off. The small stones and pebbles will be raked up in the process of cultivation, and they, too, must be removed from the garden site, for they will interfere with the cultivation of the soil and will make the preparation of a seed bed almost impossible. It is widely believed around our part of the country that stones grow, and it seems in truth as if this were so, for every year a new crop of them appears. Actually it is the action of the frost which brings them to the surface, and so the crop of stones must be harvested each year. Fortunately, each year the size of the crop diminishes. Not every gardener will be plagued with stones, but many of us are, and it is my experience that stones, large or small, or a combination of both, need not discourage one, for some of the best gardens that I know of have been grown on stony ground.

In some instances the soil from cellar excavations has been spread over the topsoil, thus rendering it unfit for the growing of anything, either lawn or garden. This soil is almost entirely mineral, wholly lacking in humus or organic material, so that there is no plant food available for vegetation. This situation can be prevented and should be if those responsible for the excavation know what they are about. Before work commences, the topsoil should be removed and stockpiled. Then, when the operation is finished, this pile of topsoil should be spread about so as to cover all the mineral soil that has been brought to the surface. If you find yourself confronted with the situation where all the subsoil has been left on the surface, there is not much you can do but buy some topsoil, or find it somewhere, and have it hauled in to cover over the mineral soil to a depth of at least three inches. Even with this expensive undertaking the cure has not entirely been effected. Deep spading and the continued addition of organic matter will be required until the depth of the topsoil reaches six inches or more.

There is a process known as "trenching" that will alleviate not only the condition we have just discussed, but any condition wherein mismanagement or abuse has resulted in a shallow and depleted topsoil. This procedure, which I fortunately have never had occasion to use, seems to be a laborious and tedious one, and my presentation of it here is purely academic, for as I say, I never have done it myself.

Actually there are two accepted procedures known respectively as "Bastard Trenching" and "True Trenching." As the accompanying diagrams indicate, Bastard Trenching is a procedure whereby the hard impervious subsoil is broken up and enriched, but it remains in its original position beneath the topsoil. On the other hand, True Trenching is employed only when there is already a good depth of rich topsoil, an effect that Bastard Trenching would tend to produce. In this latter operation the lower half of the topsoil is brought to the surface, and the surface soil is placed underneath.

In either instance the operation is slow and laborious, and if the garden is a large one perhaps the sensible thing to do, if it is necessary to break up the subsoil, is to hire a sub-soiler to do the job. A subsoil plow has a long sharp blade, with no moldboard, as I understand it, which penetrates deep and breaks up the subsoil without bringing it to the surface. I am not familiar with either these procedures or machines, never having had occasion to use them. In fact, all my experience tends to convince me that the top six inches are the important ones, and unless there is a drainage problem, there is no real need to be concerned with the subsoil.

If water stands on the ground, or watercourses wash away the topsoil, there is a drainage problem to be met. Surface water on the ground may indicate improper drainage, which is easily cured, as is too rapid surface runoff; but standing water

BASTARD TRENCHING

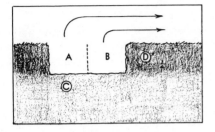

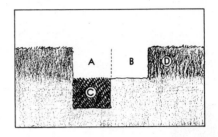

Dig trench one spade deep, two spades wide (A and B). Carry soil across garden and pile on opposite side.

At bottom left of trench (C) spade up and work in manure to depth of 8 or 9 inches.

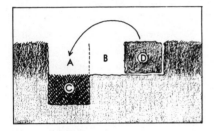

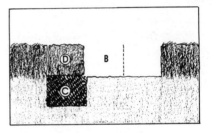

Dig new adjoining trench, one spade wide, (D) throwing soil over to cover area (C).

Repeat process from step No. 2. Continue across garden, filling last trench with transported soil from original trench. (A and B).

TRUE TRENCHING

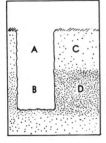
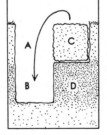
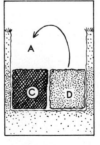
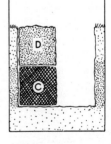

Dig trench t w o shovels full i n depth (A and B). Carry soil across garden and pile on opposite side.

Start new adjoining trench one spade deep (C). Deposit soil in bottom of first trench and mix with manure.

Deepen s e c o n d trench by o n e spade (D). Toss soil over to bring first trench up to ground level.

Repeat p r o c e s s across garden, filling last trench with transported s o i l from o r i g i n a l trench (A and B).

[15]

may also indicate the lack of proper subsurface drainage.

Now I would like to quote again from my old authority, Rawson. He says: "In treating of drainage we shall endeavor to make it clear how such course of culture operates to mellow and warm the cold barren soils, and bring them into high condition. In fact, having good exposure to begin with, by drainage, deep tilth, generous and judicious manuring . . . the most barren spot on earth can be made as highly productive as any other soil, even the richest."

Proper grading of the surface of the garden, together with garden paths and proper placing of the rows of vegetables, will prevent excessive and damaging runoff and will eliminate standing water which results from a sudden downpour.

The problem of surface water runoff will be discussed at greater length in the next chapter when we take up the subject of the placing of the rows of vegetables in the garden. Here we wish to emphasize that which is our first consideration, to wit, the water-holding properties of subsoil.

Wet, soggy soil which is the result of improper subsurface drainage will require the installation of some sort of underground drainage if the soil is to be made productive. Soils not naturally well drained derive a triple benefit from being drained. Because of deeper tilth possible on well-drained soil, the soil will retain its moisture in time of drought. The

removal of the excess water warms and aerates the soil, and, in the spring, quicker drying out will permit earlier planting. Because of its limited size, the installation of an adequate system in a kitchen garden will not be a major operation and will repay many times over the trouble and expense involved. No exact instructions can be given, for each installation will differ as the condition, size, and contour of the gardens differ.

Any one of three different types of tile may be used in building the drainage field. I have used, and prefer glazed hub tile, although there may be good reasons for using either the unglazed drain tile or perforated composition tile of the "Orangeburg" type. The unglazed tile which comes in one-foot lengths is the least expensive, but it is difficult to keep the tile in line and, lacking hubs or collars, it is subject to infiltration by sand and silt, which will eventually plug the pipe completely. The composition type pipe comes in eight-foot lengths, the sections being joined by snug collars, and may be had with perforations along the bottom. This pipe is tight and is quickly laid, but unless very carefully laid has a tendency to sag or bend, and on an eight-foot length it does not take too much of a sag to eventually plug the pipe, if there is any sediment being carried by the water. The glazed hub tile comes in two-foot lengths and is my choice as it lays easily, can be made tight and

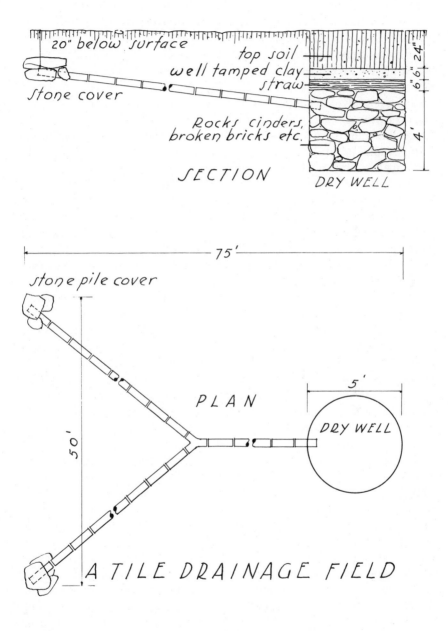

20" below surface

top soil
well tamped clay
straw

Stone cover

Rocks cinders,
broken bricks etc.

24" | 6" | 6" | 4'

SECTION

DRY WELL

75'

stone pile cover

50'

PLAN

5'

DRY WELL

A TILE DRAINAGE FIELD

it is permanent. Each of these types of tile may be had four inches in diameter which should be ample for any such project as we have in mind. The tile should be laid deep enough so that there is no danger of disturbing it in the process of preparing the soil for crops, or of being crushed if a heavy tractor be driven over it. To ensure this, the tile at its minimum depth should be at least two feet below the surface and have a continuous and even pitch of one-quarter inch for each foot. The head of the drain should be outside the garden proper and should be covered with a pile of rocks which fills the hole to the surface of the ground. The outlet must likewise be outside of the garden proper and must run to the lowest point of land available so that the drain will have free and unhampered discharge. To achieve this may present some difficulty, especially in suburban garden sites, but in these cases there is always the possibility of discharge into some already established drainage or sewage system. In my previous book, I made the suggestion that, in instances where it was difficult to find a natural point of discharge, a dry well could be built. I now consider this to be of dubious value unless a spot can be found for the construction of the dry well where there is good underground drainage, and, it seems to me now, that if this condition could be found near the garden site, there would be slim probability of any need for drainage in the garden itself. However, for what it may be worth, I am including a diagram showing a tile drainage field and details on the construction of a dry well.

In my own garden the pitch of the land is such that the discharge of the tile underground drain presents no problem. In the spring it carries off a considerable flow of water and has made possible the lightening of the soil in that portion of the garden which is underlaid with hardpan to the point where I can now successfully grow sweet corn on it. Of equal benefit is the fact that because of the drainage I can work the soil early in the spring soon after the frost has gone and, as a result, have earlier crops than most of my neighbors are able to produce.

I hold Edward H. Faulkner's book *Ploughman's Folly* in high esteem, and various conclusions that I have arrived at after forty years of gardening I find in accord with many of his, but lest some compare what I said above with his chapter entitled "Tile Treachery" and note the seeming disagreement, let me say that I propose to stick by my guns. Much of what he has to say on the subject of drainage must be considered as a theory having to do with large-scale operations whereas I am concerned simply with practice, and on a small scale at that. I agree that to drain on a large scale and at great expense may not be only foolish, but wrong as well, and I will maintain that he who tries to make a garden spot on

HOW I DRAIN MY OWN 4-PLOT GARDEN

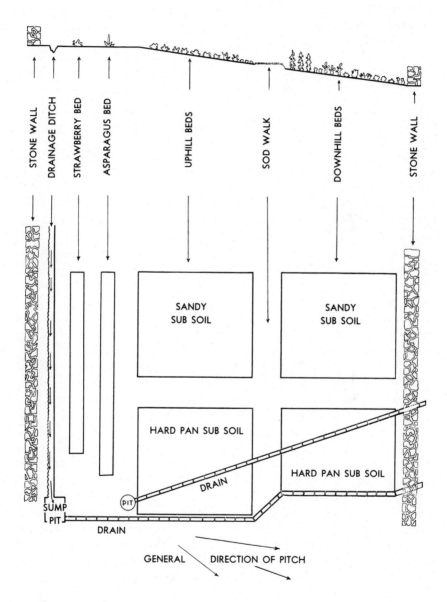

land where impervious subsoil makes the surface soil cold, heavy, and sour, and who ignores the drainage problem, will spend much of his time in struggling with the soil, and little of it in harvesting his vegetables.

At the risk of being repetitious, I want to state again that it is of the utmost importance that the soil of the garden be light and well drained. Water-soaked soil is hopeless. It is hard, lumpy, and nonfriable. It does not get enough air and, as a consequence, worms will not live in it; it turns sour, growing only moss and mold. The addition of manure and compost to such soil tends to heighten the sourness and sogginess

due to the lack of oxygen necessary for the decomposition of the organic material which has been added, whereas manure and compost added to heavy but well-drained soil will, by virtue of the breakdown of the organic matter, lighten it.

To sum up: Of primary importance in considering soils is the physical or mechanical condition thereof. By the improvement of these conditions, by draining, by removing rocks and stones, by deep tilth and the addition of organic materials, any soil can be made to be productive, even though to start with it was of the worse type, in the worst classification, and the most barren on earth.

CHAPTER THREE

A Place for the Garden

LOCATION

THE PRIMARY CONSIDERATION in choosing a site for the garden is, of course, the soil. If there is a choice of sites available, be sure to pick the one where the soil conditions are most favorable. The previous chapter will have served its purpose if it is of assistance in making this decision.

In addition to questions of drainage and stoniness, the question of present fertility is of prime importance. The nature of the growth presently on the site gives a clue as to the fertility of the soil beneath. Green and lusty growth, even though it consists most of noxious weeds, is an indication of good soil. In many localities, the worst weed the gardener has to cope with is "witch grass," also known as "quack grass," and, in my experience, the presence of this pest is a sure indication of excellent soil for garden purposes. I am certain that in many instances the prospective gardener will be seriously advised to avoid by all means any site where this baneful

grass grows. I will disagree. To be sure, bad soils can be made good, but deliberately to choose to start with inferior soil because the presence of a certain weed promises extra labor in the first stages of cultivation of the garden seems to me to indicate a lack of both energy and intelligence. To bring an inferior soil up to top condition requires season after season of careful tending, with short crops in the meanwhile, whereas to begin with good soil is of the utmost importance, even though you have a battle on your hands at first. These same timid soldiers, who are licked before the battle commences, are also sure to advise against the use of stable manures on the garden, because of the weed seeds that they carry with them. More of weeds and manures in subsequent chapters. My advice is to pick the rich soil if you have the choice, but do not let the presence of weeds deter you.

It is not possible to judge the quality or fertility of the soil by its color. While it is true that the

[21]

presence of beneficial organic materials tends to give the soil a darker hue, there are many other factors governing the color, so that to judge by color alone might well be misleading. If you will dig a hole to a depth of two feet in the prospective garden site, you will be able to see a cross section of the top two feet of the soil. The upper portion should be darker than the lower, and the depth and intensity of color of this top layer, which is known as topsoil, will be helpful guides in selecting the garden site. At the same time, the presence of garden worms is a certain indication of fertility.

This upper, darker layer of topsoil contains the organic materials which are essential to the growth of vegetation. This organic content of the soil is called humus, and it consists of the more or less decomposed remains of vegetable and animal matter; that is, plants themselves and dead animals of all kinds — mice, moles, insects, and the tiniest of microscopic organisms. Without humus, soil is dead and lifeless; it lacks the foods necessary for the growing of plants of any kind.

Organic matter, or humus, is essential in garden soils, for it serves at least four important purposes. First, it improves the texture of the soil. Second, it absorbs and holds moisture. Third, it is a host to microorganisms which break down and make available to plantlife the foods which are present in the mineral soil. Finally, it contains plant food in itself. Take in your hand some soil from a good garden or greenhouse; examine it carefully; smell it, squeeze it, rub it between your fingers. Now take a handful of bank soil or dirt from an excavation or gravel pit and give this sample the same careful examination. Such a test will convincingly demonstrate the characteristics of humus and will be helpful in choosing your garden site.

Our ultimate aim will be to have a plot that will receive and hold an adequate, yet not excessive, amount of moisture. Thus there are two factors to be kept in mind: natural drainage and protection from winds

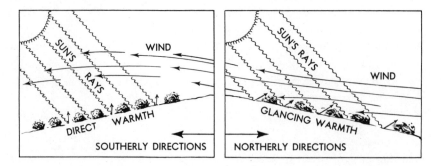

Southern slopes give maximum warmth and protection from wind.

[22]

which might cause rapid evaporation. At the same time, we will want the maximum of direct sunlight. The search for surface drainage, protection from the wind, and exposure to the direct rays of the sun will lead us to a site which has a gentle pitch or slope.

In this part of the country the ideal condition would be to have a southern or southeastern slope, for such a pitch faces most directly into the sun and affords protection from our prevailing winds from the northwest. In places where the prevailing winds are strong and unrelenting and the summer season is short, these considerations are of prime importance. Country dwellers in the north can testify that snow lingers on northern and western slopes long after southern and eastern slopes are bare and warm.

Planting on a southern slope will necessitate the rows of vegetables running east and west, for it is imperative that the direction of the rows be at right angles to that of the slope. Rows planted along the line of the slope, rather than at right angles to it, act as channels for the runoff of rain water, to the extent that serious loss of topsoil will result. On this important subject, more later. There is a theoretical advantage accruing from having the rows run north and south; under these conditions the sunshine falls in the direction of the rows and there will be a minimum of shadow cast by one row upon another. To my way of thinking, this factor is of minimum importance and is far outweighed in fact by considerations of warmth of exposure and protection from prevailing winds. If the only available garden site is on flat land, the slight advantage of north-south rows will be worth taking into consideration.

In any event, whether on sloping ground or flat, keep it in mind to take advantage of any existing windbreak possible. Such a windbreak may be a hedge, a row of trees, a wall or building, but in no case should the garden site be close enough so that any part of it lies in the shadows cast, or so that encroaching roots rob the garden plot of its fertility. In general, we are looking for a warm and protected spot where the sun lies long and the shadows are brief and fleeting.

The ideal is not always obtainable; in fact, more often than not, we must accept conditions not of our own choosing. A case in point is my own garden which slopes to the north, has its rows running from east to west, and has not natural windbreak against a strong prevailing northwest wind. In spite of this, it seems to do very well indeed.

Finally, there is one other point to be considered in selecting the site for our garden: let it be as conveniently close at hand as possible. It is always easy to sneak in those few extra minutes which are so important if the garden is right there for you to step into. Furthermore, if you live in the country, remember

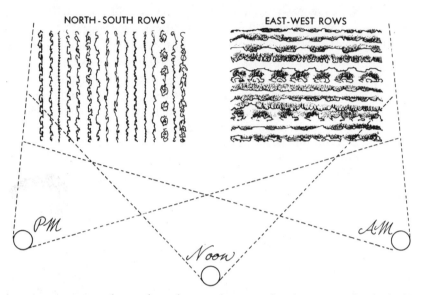

NORTH-SOUTH ROWS   EAST-WEST ROWS

A common argument for north-south rows, demonstrating that sun reaches both sides of plants daily. True enough, however . . . see text.

the rabbits, skunks, coons, woodchucks, and other marauders of the woods and fields. They may prove to be a continuing problem, but if the garden is close to the stir and bustle of domestic activity, there will be less chance of damage from wild animals.

While other gardens in the vicinity are suffering from the depredations of coon, deer, and other wild connoisseurs of succulent vegetables, my garden has never been damaged in the least. I can only conclude that this is because it is close to the house, which is the center of con-

Garden plants cannot compete with tree roots for moisture and food.

siderable activity on the part of dogs, children, and household comings and goings. I might add that my two bird dogs are trained to keep out of the garden.

Summing up what we have discussed thus far, in choosing the garden site we must consider the following:

1. The existing fertility of the soil.

2. The pitch of the land with its subsequent advantages of drainage, wind protection, and exposure to sunlight.

3. The proximity to existing windbreaks, taking into consideration the necessity for avoiding undue shade and the competition of the long roots of trees, shrubs, etc.

4. The accessibility to the house and protection from marauding wildlife.

Admitting the improbability of locating the ideal spot which embodies all of these advantages, let us not compromise with the most important factors, which are fertility and drainage, with proximity to the house running a close second.

A study of the photograph of my garden will reveal to what extent I have put into practice my own preaching. It lies on a gentle slope (although the direction of the slope is not ideal); the rows run counter to the slope, forming miniature dams; the garden is divided into plots by strips of lawn; a wall at the north end serves as a windbreak and as a stop against excessive erosion;

it lies close to the house; it is free from the shade of trees or buildings, and has no competition from the roots of shrubs or trees. Its location is the result of a series of compromises, but it is the best I could do with what I had to work with. However, its productivity is such that I feel no need to make any apologies whatever.

Actually, in planning and growing a garden there is no one way which can be said to be the best. Each garden presents a different problem, which, in each case, is a challenge. There is no satisfaction greater than that of obtaining the maximum yield from a given plot of land. It is the purpose of this book to show the results one may expect from the amount of time and labor spent there.

SIZE

Having decided upon the location, the size of the plot is determined by three important factors: the amount of usable land available, the size of the family the garden is to serve, and the amount of time that can be devoted to garden work.

If only a small plot is available, there must be careful consideration of the kinds of vegetables to be planted and those which can be dispensed with. Potatoes will be eliminated, for they require a great deal of space; so, too, do melons, winter squash, and pumpkins. If no one in the family likes parsnips, for-

get them. Staple winter vegetables, such as turnips and celery, which can be purchased reasonably at any greengrocer, may likewise be omitted.

In starting a garden with no fund of experience to draw upon, one might be bewildered by the seeming problem of size. Actually, there is no problem. If you have a space in your backyard twenty or thirty feet square, you will have room enough for a serviceable and worthwhile garden. If you are fortunate enough to have unlimited space, there is still no problem. You need only bear in mind the size of your family and the amount of time that you or others can devote to work in the garden.

It is both unwise and disheartening to attempt to care for more garden space than can be properly attended to. Comparatively little produce, usually of inferior quality, will result from an ill-kept garden. Nothing is more dismal and discouraging than a plot of weeds whose vigorous growth has almost completely obliterated all traces of vegetables. A stunted carrot here and there, bullet-like beets, scarcely larger than a marble, a few pale and sickly lettuce leaves, these are the rewards of the gardener who lets the weeds get the better of him.

Weeds will be discussed later, but let everyone who contemplates a vegetable garden take them into consideration and realize that no useful species will grow in competition with weeds, and that it takes time and trouble to control the weeds.

At this point do not consider the use of weed-killers to minimize the amount of time and labor involved, for not only are the results questionable, but the effects on the soil and the general ecology of the ground treated, is certain to be harmful. Nor should the gardener think that there is some magic he can invoke to help in the battle with weeds. I shall go into this in more detail later, but let me say right here that the hope that there can be any "green thumb without an aching back" is a will-o'-the-wisp. So, in making the preliminary decisions as to the size of the garden, full consideration must be given to the time necessary to have the plot well weeded and otherwise well cared for. On the other hand, it is amazing how much a small area of intensely cultivated garden will produce. Always bear in mind that intensive cultivation is more important than extensive cultivation.

There is prevalent in agricultural circles in this country the conviction that all the solutions to problems of production can be found in the increase of production in terms of the number of man-hours involved. This formula has worked as far as increasing commercial production figures is concerned, but there remains some question as to the ultimate validity of this approach. At any rate, the proper approach for the worker in the kitchen garden is not production per man-hour, but production per unit of land. As far as home gardens are concerned, intensive cultivation is the proper way to go about things.

To sum up, the size will be determined by family needs, space avail-able, and the amount of time which can be devoted to the care of the garden. Whatever you estimate this size to be, it is better by far to underestimate than to overestimate, for to be forced to abandon a part of the project may result in loss of enthusiasm for the whole; on the other hand, it is easy enough to enlarge the garden from year to year as needs are determined and competence increases. I realize that such broad generalities as the above may be of slight value in helping to solve the problem of the rank beginner who has had no experience in growing vegetables, but I am loath to state too explicitly any rule as to size.

In succeeding chapters the details of preparing the soil, planting, and cultivation will be given, including detailed discussion of the various vegetables most likely to be used in the family vegetable garden, together with yields to be expected, amounts of seed necessary, etc. It would also be helpful to have the following information:

Amount of seed necessary for one hundred feet of row.

Proper spacing of rows and plants.

Number of seeds per ounce.

Amount of seed for given number of plants.

Dates for planting.

Depth for planting.

Days to harvest.

Using the information to the best advantage each one will have to make his own decision as to size, keeping in mind always that it is better that the plot be too small than too large.

PLAN

The location and size of the garden having been determined, the next problem to consider is the layout or plan thereof. In general, each vegetable garden, even a small one, should consist of two or more separate plots, each part divided from the other by a strip of greensward. There are good reasons for dividing the garden site into plots in this manner.

Any garden placed on a slope will be subject to soil erosion. In addition to wind erosion, which good management can hold to a mini-mum, there is bound to be in the nature of things a movement of the soil under the influence of water, from the higher parts toward the lower. The larger the plot whereon this movement goes on unchecked, the greater the amount of soil that is transported. In other words, on a slope the upper reaches are slowly but surely being deprived of their topsoil, while on the lower parts, if means are provided for the precipitation of the silt, the depth of the topsoil is increased. The principle involved is the same as that of the sedimentation of dams. The accompanying diagram will perhaps clarify this point.

In the agriculture of the Old World, in Italy and Spain, in Asia, India, and Japan, and in the Andes

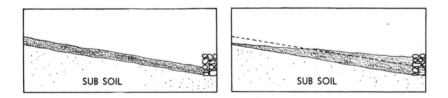

(A) Topsoil worked and loosened to uniform depth on sloping ground, as at left, will eventually move downhill, as at right.

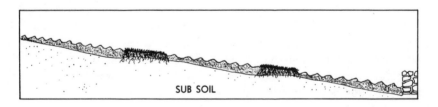

(B) Downhill movement of topsoil kept in check by sod strips across the slope, and a wall or dam along the lower border. Planting of rows cross-slope also checks excessive run-off.

[28]

of South America, terraced gardens are seen wherever the terrain is hilly. Following the contour of the hill and separated by narrow strips are a series of stone walls which hold the soil in place. Having chosen a garden on a slope, we must accept the necessity for holding our soil in place, and the means of accomplishing this will be a sod strip or strips across the contour, with a wall or a dam of some sort at the very bottom of the garden. When sudden summer cloudbursts descend, the action of these strips or dams is immediately apparent.

Besides this important function, separate plots permit the principle of crop rotation to be carried out to its fullest extent, for in this case the soil from one plot is not dragged into another by the business of plowing or cultivating; a rotation, an important objective of which is to avoid soil-borne diseases, therefore has a much better chance for success. The principles of crop rotation go further than simply the control of soil-borne diseases, but in this single aspect, separate plots are an important factor and, in general, the whole operation is simplified by their use.

In any discussion of crop rotation the general groupings are more or less arbitrary. In my garden, which has four plots, I have roughly four separate groups which are planted each year in successively different plots. I have no hard and fast rule except that there shall be a complete change each year, with particular pains taken to see that the Brassica group (that is, the cabbage family) will have at least a two-year break in its rotation. For the rest, the corn has one whole quarter each year and is moved among three of the plots; peas, beans, and some small seeded vegetables may be together in another plot, with all of the transplants plus squash in a third plot, and all of the small seeds plus, perhaps, cucumbers in the fourth plot.

The general principle is not dependent upon the number of plots, however, and beneficial effects will result if only two plots are used. There is no inviolable rule for the alternation of crops that I know of. Rawson says to alternate deep-rooted crops with shallow and slow-growing with those of rapid growth. No root crop should follow a root crop, nor should vines follow vines.

In general, crops differing widely in their food requirements should follow one another. Leaf crops remove nitrogen from the soil, the legumes replace it. Root crops require abundant potash and seed crops demand ample supplies of phosphorous. From these general principles, and from the depths of one's experience, each will work out the plan of rotation which best suits the type and size of his undertaking. More details of crop rotation and the necessity for it will be discussed in a later chapter.

If we are committed now to a garden consisting of two or more

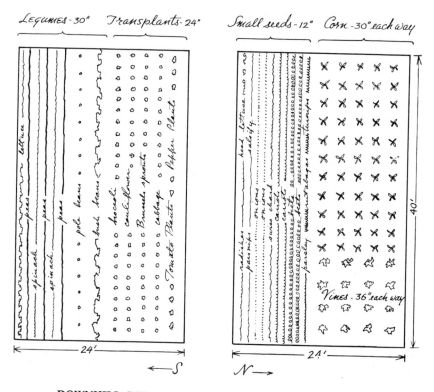

DOWNHILL BED

UPHILL BED

A two-plot plan, average in size, in which the garden is separated into groups that have equally spaced rows. This looks convenient, but may not be when you come to plant. (See Chapter 6, page 73.)

plots, let us proceed to make a plan for the garden and to put that plan on paper. I must confess that after forty years of gardening on the same plot I no longer draw out a plan for my garden, but for the beginner it should prove useful to do so. First, decide upon which vegetables you wish to grow; of great assistance in this process is the use of a seed catalog from one of the large and well-established seed growing firms. Having made this first step, procure a piece of smooth brown wrapping paper, twenty-four by thirty inches or larger, and, with a ruler or scale, lay out the garden plot using a scale of one-half inch to the foot, or, if you can make room, it would be less pinched to use a scale of one inch to the foot. By the same token, reduce the scale to one-quarter inch per foot if you are cramped for room. The plots having been charted in position, we are now ready for the next step.

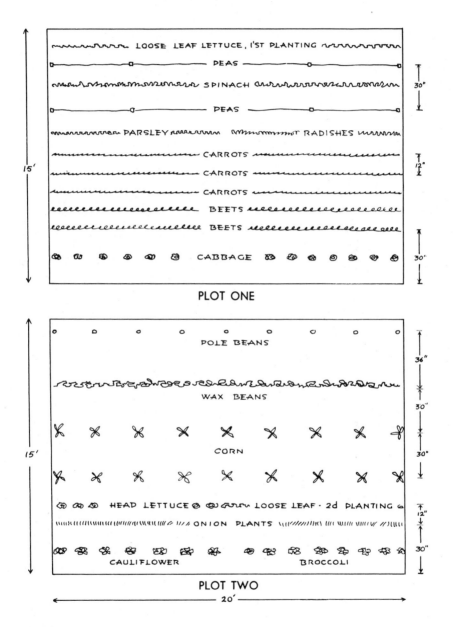

A small double plot, each 15' x 20'. This is a good one for a beginner. A plan quite practical for orderly preparation of the seedbed. (See Chapter 6, page 72.)

[31]

There are several considerations to be kept in mind at this point; first, to group the plantings in the separate plots according to the requirements of crop rotation, and then to contrive to have such vegetables as may be planted at the same time kept together as much as possible. For convenience, it is well to have the transplants kept together.

A glance at the charts showing the groupings of vegetables in the four different plots of my own garden in the past two years will give a graphic idea of the principles involved.

Speaking of garden lay-outs, some authorities raise the question of compatability of crops. I have no quarrel with those who hold, for instance, that hemp will repel the cabbage butterfly, or that marigolds will keep bugs away because of their smell. But I must say that for my part, I have never seen either favorable or adverse results which derive from the association of plants. My suggestions are strictly practical and have to do with the time to maturity, the cultivating requirements and such.

Having decided in general which

MY GARDEN PLAN, ONE YEAR

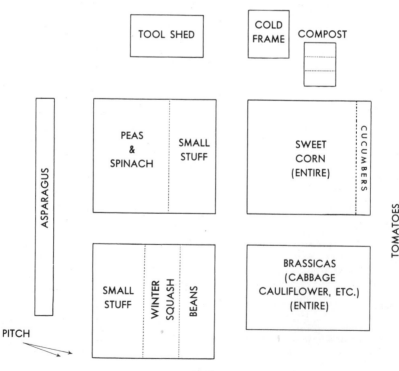

MY GARDEN PLAN, THE FOLLOWING YEAR

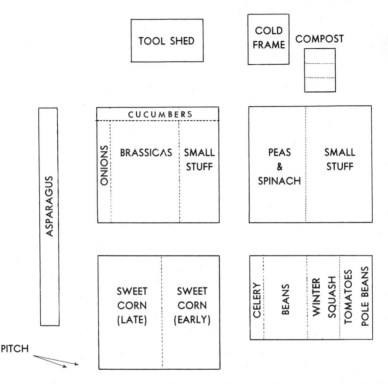

vegetables are to be planted in which plots, we can proceed with laying out the garden plan. Now, using the scale, lay out the rows of vegetables. In order to do this, it will be necessary to know how far apart the rows are to be. Different vegetables require different spacing, but the simplest and therefore the best plan is one which will involve us in the least number of variations in the width of rows. Each packet of seeds will have planting instructions on the outside, but in the search for simplicity perhaps some of these directions can be modified.

I take it to be a matter of plain common sense that if all of our energies are expended in the direction of getting the garden soil up to its optimum condition, it is wasteful to allow large strips of it to lie fallow between the rows. Thus it follows that all rows will be as close together as will still permit proper conditions for the growth of the plants. This means that allowing wide spacings between rows for wheeled tools of one sort or another is wasteful, and not to be undertaken. The same reasons lead to the decision that in most instances the extra space re-

[33]

quired if the plants are to be mulched, is again a waste not to be tolerated. More of mulching and power cultivation later, but for the small kitchen garden it can easily be demonstrated that the urge to avoid labor which results in wide spacing, actually accomplishes exactly the opposite of that which is intended.

For these reasons I recommend that all small seed be planted in rows spaced from eight to ten inches apart, peas in rows thirty inches apart, corn in hills thirty inches apart in each direction, that most of the transplants can be placed eighteen inches apart in rows twenty-four inches apart, summer squash in hills three feet apart in each direction, and beans in rows thirty inches apart. Let us also assume that we have decided on a small two-plot garden, each one of which is fifteen by twenty feet. You may end up with a plan more or less like one of the several shown here. However, it will probably not be possible to come up with a plan which will enable you to complete planting in one plot before you will have to start planting in one of the others. Of this problem we will speak at greater length in Chapter Six.

You will note that early vegetables, which are removed from the garden as they are used, can be planted between the rows of peas. Radishes, loose leaf lettuce, and spinach can be planted thus, for they will be out of the way before the peas are high enough to shade them overmuch. In localities where there is a long growing season, succession plantings can be planned as well; but for the beginner, it will be well to stick to a simple one-planting plan at first.

With a tentative plan now on paper, we have something to go on and, by balancing the space available against the amounts of each variety we want, we can finally arrive at a layout which will show how many rows or hills of each there will be. Applying these figures to the tables in the appendix we can ascertain just how much of each kind of seed we must buy. Let me give fair warning: Unless you exercise restraint you are apt to end up with more summer squash, and possibly string beans, than you really want or need.

In any event, let us assume that mistakes will be made. Real wisdom in this department will come as a result of trial and error. It will be impossible to foresee all problems or to estimate all needs accurately, even though we could count on Mother Nature to behave exactly on plan and schedule.

[34]

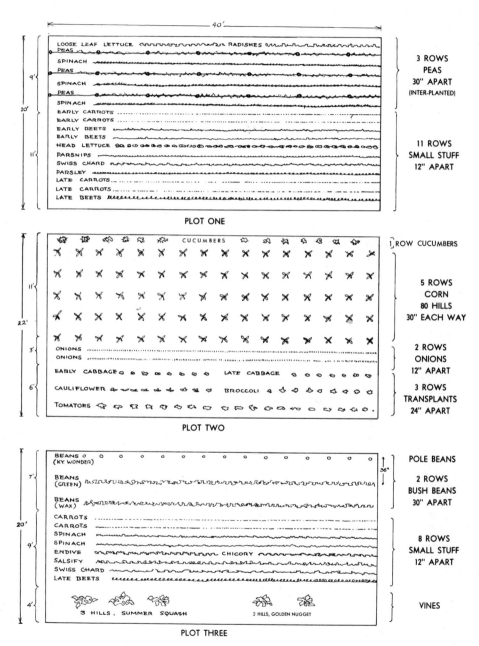

PLOT ONE

PLOT TWO

PLOT THREE

A three-plot layout, somewhat more ambitious than the others suggested. Mine is much bigger, but it feeds a whole neighborhood.

CHAPTER FOUR

Food for the Garden

BEFORE THE GARDEN plot can be planted or even, for that matter, before the seedbed can be prepared, the soil must be treated. That is, material must be added so that the fertility is increased, and perhaps humus should also be added. This is a process which must be carried on from year to year, for as plants grow they remove certain substances from the soil. In addition, other factors of climate and weather tend to subtract from the store of available plant foods. The whole subject of the treatment of soils is a very complicated one, and it includes problems of moisture content, texture, and humus supply, as well as those of plant nutriments.

A previous chapter on soils discussed some of these problems, and it may seem repetitious to discuss the same subjects in this chapter. This is not the case. When plant foods in one form or another are added, the physical structure of the soil is altered in the process, so while de-stoning and draining form important steps, they are, for the most part, but preliminary ones. What happens to the soil in our garden after it begins to produce vegetables for us is our responsibility and is a result of the manner in which we handle it. Proper treatment will not only maintain the fertility with which we started but will improve conditions as well, so that our yields will increase from year to year until the maximum has been reached. How to achieve this desired end is a subject which has filled many volumes. Very few authorities that I have come across are in complete agreement, but in general they may be placed in either one of two diametrically opposed camps.

On one side we find the disciples of the scientific doctrine; on the other side are the disciples of the organic doctrine. On the extreme fringe of the scientific doctrinaires are those who would grow plants in chemical solutions, kill bugs by supersonic vibrations, inject vitamins by radio-active products of atomic fission, and in the end nourish the population of the world on a diet of pills. Their counterparts in the organic camp shudder at the very thought of a chemical. They treat their soil with compost made in piles of mystical proportions and occult dimensions, piles which have been treated with magical solutions whose components are not to be revealed to the uninitiated. To put it less flippantly, the scientific school is concerned primarily with producing growth by feeding the plants through their roots with soluble chemicals and by controlling weeds, pests, and disease by the use of chemical dusts and sprays. They are interested very little, if at all, in the soil itself. To get the most out of the soil with the least effort and expense is the primary objective, and that seems to be, on the face of it, a reasonable and practical one.

The organic gardeners, however, disagree violently with this procedure and its objectives. Their method is to study and understand the whole cycle of life in the garden, both in the air above it and in the ground beneath it, to the end that the organic whole may function harmoniously in nature and as nature intended, carefully guided by the hand of man to his useful ends. Their scope includes birds and airborne insects, spores, and pollens, as well as the more important soil itself, with its myriad insects, worms, fungi, and bacteria. Their objective is so to plant, fertilize, and cultivate that this state of organic balance may be achieved, for then there will result the maximum production of vigorous, succulent, and health-giving vegetables.

Starting off on my own some forty years ago with the tradition of old-fashioned home gardening, as practiced on a self-contained farm unit, behind me, I found that many of the gardening methods which came down to me from my grandfather were similar to those which I subsequently discovered were advocated by the organic gardeners. However in those days one did not hear much about organic gardening. In fact, all the information available on the subject of gardening was issued by the other camp, through Department of Agriculture Bulletins, county agents, agricultural periodicals, etc. This unrelenting flood of scientific dogma had its effect; therefore, in 1942, when I wrote *How to Grow Food for Your Family*, I said some things in that book, even though there were faint doubts in my mind at the time, which resulted directly from having heard and read so much of the scientific point of view. I had ever been an active conservationist, a lover of wildlife, and a firm believer in "balance in nature," but the volume of propaganda offered by the agricultural scientists was so great that I failed to correlate what I had observed in the wider field of nature with that which was taking place in my own garden. As a result, I found myself in a position which was not well thought out, and naturally this ambiguity revealed itself in what I wrote.

I recognize now that there can be no compromise between the two divergent points of view, and one must take a stand in one camp or the other. The great fundamental truths of conservation, if they apply to fields and forests, must apply to gardens as well. If silted and polluted streams, lower ground-water tables, dust bowls, and eroded fields result from man's greedy defiance of natural laws, then we must expect to find these same evils duplicated on a much smaller scale in our gardens if we persist in our defiance.

This is the day of concessions, compromise, and "peaceful co-existence," so that now it is sometimes difficult, if not impossible, to decide whether the "authority" belongs to the one camp or the other. For example, take this excerpt from the "Country Correspondence" column by the late Hayden Pearson: "The only reason that I recommend large quantities of fertilizers is that it means more profits. . . . It is a double-barreled program: soil in fine structural condition, and soil with plenty of chemicals to grow big crops." Intellectually one accepts the necessity for humus in nature's cycle of growth and decay, but the necessity for quick profits dictates the addition of large quantities of chemicals, which in turn destroy the humus. As I have said, I see no need for compromise, at any rate not in a home garden.

For over thirty years now I have run my garden without the use of any chemical fertilizers or sprays, and the results have exceeded anything produced by the garden in the previous years, during which period I did use them. So it is that I take my place on the side of those who eschew the use of chemicals, and what I will have to say on the subject of soil conditioning and fertilizing will not include anything on chemical fertilizers, dusts, sprays, or weedkillers. My approach will not be that of little labor and prodigious profits, but will be in the direction of soil betterment to the end that the garden may produce to its maximum from year to year without making any drafts against our capital.

It should be noted here that there are several commercial preparations for use as soil conditioners and insecticides which are not properly in the category of chemicals. In the field of fertilizers the list would include bone meal, fish meal, ground limestone, ground phosphate rock, sphagnum moss, etc. Among the insecticides there are rotenone and pyrethrum, the latter being prepared from the dried flowers of three species of chrysanthemum, while rotenone is prepared from the roots of two tropical plants, derris and cube.

While on the subject, it will be well to note that within the past few years a new preparation has come on the market, loudly acclaimed and widely heralded as the complete and final solution to all the problems of modern agriculture. This material, which has appeared under several different trade names, is an agent which, when mixed with the soil, is supposed to loosen and lighten the heaviest and soggiest of them. In these preparations we find the scientist's solution of a scientific dilemma. That the continued applications of chemical fertilizers destroyed the humus was a conclusion which could not be avoided, and while the necessary plant foods were being supplied, the condition of the soil deteriorated to the point where the all-important condition of looseness and friability was lost.

Not far from this place, on the right of the road as one proceeded down along the West River, there was a narrow strip of elegant alluvial soil. Each successive year for ten or twelve years this one acre plot was planted to potatoes, and for the first few years the growth and yield were excellent. That which I observed taking place here over the years has not been checked out with the potato grower, I am only reporting what I saw as I drove past the place. Soon the stand of vines became spotty; with each succeeding year thereafter showing a decided lessening of healthy growth. Here I assume (an assumption supported by the presence of fertilizer bags by the edge of the plot) that commercial fertilizers were added in ever increasing amounts as the years passed.

Finally the potato grower admitted defeat, and he quit that which was obviously a losing game and the land lay there fallow. But the interesting thing about it is that for five succeeding years not even weeds would grow on the plot. Scrubby grasses vied with the moss to cover with some sort of decent sheet the violated and naked ground. I kept no record of events, so I cannot give actual dates, but this year, I guess some six or eight years after the potato grower gave up, an attempt was made to grow a vegetable garden on the spot. Let me assure you that no more pitiful display of dwarfed plants and gaping rows could be found anywhere.

My observations here are certainly not the result of scientifically documented facts, but I will let them stand. He who robs the soil will destroy it; he who feeds the soil will cause it to multiply ever more vigorously from year to year as my garden has done. Here the logical conclusion would be that there was something wrong with the method, but since this would discredit all the work of the chemical agriculturists from Lebig on down, such a conclusion must be rejected. To be right, a scientific solution to the problem must be found, and this chemically inert mechanical soil loosener is the result. In following the path of science in the treatment of soils we are led into an ever-increasing expense for fertilizers, as in each succeeding year greater quantities of chemicals must be added to achieve the same result, and then in the end we come to the final expense of repairing the mechanical damage which these chemicals have caused, by adding another and more costly material. What the net result of this procedure will be cannot be predicted, but I suspect it will end in a stalemate as the enthusiasm for these mechanical soil conditioners seems to have died down within a year or two of their introduction to the public.

The ironic part of the whole procedure is that he who works the soil using nature's method will discover that, as the fertility improves, so does the mechanical condition, and vice versa. There is an epoch-making book on the subject of nature's methods of soil management — Sir Albert Howard's *An Agricultural Testament*. If I could have my way, every intelligent person would read this book, not merely those who live on the land and work the soil, or who hope to raise vegetables in their gardens.

Let me make it clear that I do not maintain that moderate applications of chemical fertilizers will ruin a soil which is kept rich in humus. What I do maintain is that continued applications of chemicals, and chemicals only, *will* ruin the soil. I will grant that slight damages to the soil structure may be repaired by continually adding humus, but the use of chemicals in almost any degree is apt to interfere with nature's operations.

And the dislocation of nature's cycle is almost certain to show up in susceptibility to disease and insect damage on the part of the vegetables.

Then there is a final clincher to the argument, and that is that vegetables grown without the use of chemicals not only taste better but are more healthgiving. On this last subject there is much being written, some of it so faddistic in tone that it may tend to prejudice the conservative reader against all such mumbo jumbo; but before you discard the whole business as the brainstorm of a bunch of faddists, read the chapter entitled "Soil Fertility and National Health" in Sir Albert Howard's book mentioned here. As to the difference in taste, there is no room for controversy; I am sure that once you make the test you will be convinced.

On the general subject of soil conditioning by the use of manures and compost exclusively, there is another conclusion which I have arrived at after years of observation, and that is that the question of soil acidity or alkalinity ceases to exist for all practical purposes for the vegetable garden.

It is my experience that the growing requirements of all the common garden vegetables are so generally similar, as far as acidity or alkalinity are concerned, that all will thrive and prosper equally well in a garden where the soil is in a high condition of organic perfection. The classic example of incompatibility given by

BUILDING THE COMPOST PILE

1. Lay out rectangle 5′ x 12′.

2. Lay up sods for outside wall.

3. Garbage goes in daily. Cover lightly with earth or manure.

4. Add garden waste throughout summer. Build sod walls higher.

5. At season's end, pile is roofed over with inverted sods or manure.

1st WINTER   SUMMER   2nd WINTER

6. Leave it alone for a year and a half. Let nature do the work.

[41]

the experts is that, if spinach does not do well, the soil is too alkaline, and if beets do not do well, the soil is too acid. The inference is obvious, and that is that both beets and spinach could not possibly do their best in the same type of soil. On the basis of actual experience I cannot agree with this conclusion, for in my garden they both seem to achieve their maximum potential. If this be true, then much that is said on the necessity for liming our garden soils — and it is being continually dinned in our ears — can be ignored as being unimportant.

There can be no doubt but that in some instances liming is important, for many soils which are poorly drained or lacking in humus will be too acid for any growth except that of plants or shrubs most tolerant to acidity, such as laurel or blueberries. It is also probable that in some instances the processes of nature, starting out with a mineral soil derived from acidic rocks, will build up a topsoil which is too acid in reaction to be good for garden use. These soils certainly should be treated at the start so that the excessive acidity is neutralized, and the use of agricultural lime is recommended. Once the initial overacidity is eliminated, however, and, by the use of manures and compost, the soil is brought to a proper state of organic balance and mixed crop rotations are established, the need for continuous liming of the soil will cease to exist.

This last statement is so heretical I must admit that it took courage to write it. Nevertheless, for the past thirty years I have added no lime to my garden, and yet I need to apologize to no one as to the results. It is true that I do use wood ashes, but I use this material for a specific purpose of which I will speak at greater length later. In general, wood ashes are alkaline and, if fresh, contain soluble potassium carbonate; therefore, the use of wood ashes would tend to neutralize acidity and to add valuable potassium to the soil.

However, my use of ashes has been neither specifically to neutralize nor to add potassium. I sidedress the cabbage family with wood ashes to control club root and maggots; the turnips and onions are sidedressed for maggots, and the beets for scab. This material is applied locally and not broadcast; and it may well be that the beneficial results which accrue from its use, as I have indicated, are not directly derived from its action on the pests and diseases, but indirectly in resistance built up in the plants by feeding them the little different something they need. Even if the latter be the case, it seems to be a better system than to broadcast neutralizing materials indiscriminately on a garden plot wherein the growth shows no general need for such treatment.

My conclusion then is that a garden soil which is in the proper state of organic balance will have no need of soil analysis or the specific addition of nitrogen, phosphorous, potash, lime, or chemical soil conditioners. To keep it in this proper state, however, does require the continual and liberal addition of manures and compost, and I will give here these procedures as I have used them.

Going back to the kitchen gardens of the self-contained farms of our grandparents, we discover that conditions and procedures pretty much parallel those set up as ideal by Sir Albert Howard. There was no large scale monoculture, there was plenty of manure of different kinds on the place, no chemicals were used, and crops were rotated. Most significant was the fact that the standard of production was per unit of land, not per man-hours of labor. Composting was seldom resorted to for there was a plentiful supply of well-rotted mixed stable manure available for use on the garden plot. A suitable plot of good loamy soil was selected near the house, it was liberally covered with manure in the fall, turned under and harrowed in the spring, and the garden planted. It was as simple as that, and it still is, with one big exception: almost none of us has access to unlimited supplies of well-rotted stable manure.

I'm afraid there is another big difference which is significant, and that is that we all accept without question a standard of production based on man hours of labor, rather than one based on units of land under cultivation. The difference be-

tween the two standards is greater than is at first apparent, and it must be understood and appreciated if we are to get on the right track. The first of these two objectives is that of the scientific point of view; namely, "to get the most out of the soil with the least effort and expense." The second is that of the organic gardener whose aim it is to bring each unit of soil up to its highest possible state of sustained production. This latter standard has been accepted for the most part by all civilizations in the past whose economy was or still is based on agriculture, where access to new lands was limited, and where farms were handed down from father to son for generation after generation. In these cases soils never became "worn out" with use; in fact, they improved with use, and the best lands were those longest under cultivation. This condition can still be found to exist in Japan, for instance, where the best lands are the old ones, and the poorest ones those most recently reclaimed or transformed from wild states to cultivation.

With the mechanization of society, with an economy based on industry, this standard became outmoded, and "farm factories" devoted to monoculture with chemical feeding and mechanized cultivation came into being. Here, success in any given year had to be measured in terms of man hours of labor. That this point of view is responsible for the loss of fertility in the lovely Shenandoah Valley, for the dust storms in the West, and for the tremendous loss of topsoil by erosion, there can be no doubt in the minds of those who have studied the problem. To what end its continued acceptance will lead us is a matter of speculation and far beyond the scope of this book. For us, who at the moment are interested in making a good garden, the greater problem can be ignored for the time being, at least, but we must come to grips with the lesser one, and that is, where and how to hold enough organic material for our needs.

Our first search is for manure. In passing, it is interesting to note that the word "manure" comes from the old French word "manoeuvrer," which means to cultivate by manual labor, so it is strictly an agricultural word and one that ties in directly with our theories of organic gardening. Manure today means stable, hen house and barnyard droppings, mixed with bedding of various sorts, straw, old hay, sawdust, or shavings. No real farmer will sell his manure at any price, for it must all be returned to the soil, so we will have to seek elsewhere to procure it. Poultry establishments may have hen manure for sale, and riding academies, horse manure; mushroom farms will sell the rotted manure from their beds which can be used only once. There are other occasional sources of supply, stockyards, for instance, or logging operations where horses are used.

Nowadays bagged manure of various sorts is available commercially, but it is very expensive, too much so for general use, I'm afraid. If you are so situated that you can buy manure reasonably and in sufficient quantities, you are fortunate. Last year I bought stable manure at four dollars a cord (a cord is a pile four feet wide, four feet high, and eight feet long) plus transportation. For a couple of years I had mushroom plant manure delivered eighteen miles at about twenty dollars a cord. This price was out of line for me as other satisfactory manure was available at a lower price, but it is not too high for garden use in general.

Fresh manures are not, in my opinion, desirable for garden use; they should be rotted or composted first. If the supply comes from old piles, this process has already taken place, although possibly not in the ideal manner, for manure in piles will "burn," thus losing a great deal of its goodness; or it may stand with "wet feet," and the wetness, by excluding sufficient oxygen, will turn the pile into a black, sour, soggy mess, also of inferior value as organic material. Raw manure should be piled, sprinkled with topsoil, and occasionally turned with a fork. Poultry manure should be mixed with at least its own volume of topsoil before it is ready for garden use, unless it be spread thinly in the fall and turned under in the spring. Phosphate rock, ground lime-

MULTIPLE PILE SYSTEM, STEP BY STEP

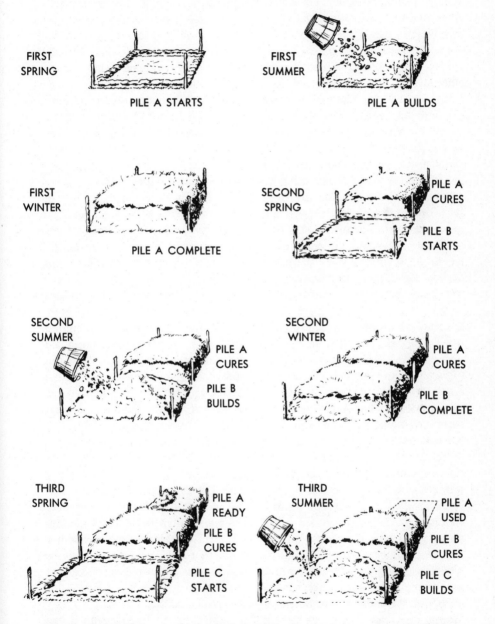

FIRST
SPRING

PILE A STARTS

FIRST
SUMMER

PILE A BUILDS

FIRST
WINTER

PILE A COMPLETE

SECOND
SPRING

PILE A
CURES

PILE B
STARTS

SECOND
SUMMER

PILE A
CURES

PILE B
BUILDS

SECOND
WINTER

PILE A
CURES

PILE B
COMPLETE

THIRD
SPRING

PILE A
READY

PILE B
CURES

PILE C
STARTS

THIRD
SUMMER

PILE A
USED

PILE B
CURES

PILE C
BUILDS

I have called it the lazy man's method, but it is also the busy man's method. It's a swap. Longer elapsed time for shorter working time.

[45]

stone, bone or fish meal may be mixed in with the composting manure for the purpose of supplying additional quantities of phosphorous, calcium or nitrogen.

In my experience, well-rotted manure will need nothing added to make it suitable for garden use. Well-rotted stable manure will be odorless, an even, dark brown in color, and will be crumbly in texture. If you can get sufficient quantities of this, or if you can get the green manure and make it, you are all set for your garden. If you cannot, then you will have to resort to a compost pile. In any event, a compost pile is a sensible and even necessary adjunct to a garden, for it means the conservation of waste and a reduction in the expense of operation.

The difference between manure and compost is simple: in the first instance, an animal feeds on vegetation and passes the material through its body, extracting nourishment in the process. Thus the waste consists of organic material which has been fragmented and treated with body juices, then subjected to further decomposition due to the complicated action of oxygen and bacteria while the manure is stacked in piles. Compost is, in general, made in the same way, with the exception that one step in the process is omitted, that of passing through the body of some animal. The end products are highly similar and, for our purposes, nearly identical. There are several different variations of these procedures. Rather than to attempt to give details of all of these methods, I will give only my own, with references to published material wherein other procedures, and possibly better ones, may be studied.

I start in the spring by laying out on a level piece of well-drained ground a rectangle about five feet by twelve feet, marking the corners with stakes. Then I lay up an outside wall of one or two thicknesses of sod or cement blocks. My system requires the maintenance of two compost piles, one of which ages for a year while the other one is being built, so in preparing for current use the pile which has stood a year, I strip it of all outside material, much of which is only partly decomposed, and place it within the borders of my sod strips as the first layer in my new compost pile. From now on all decomposable garbage from our house, and from our neighbor's as well, if I can get them to sort their waste, is spread on the pile and covered with a thin layer of topsoil before it has a chance to become nasty.

Early in the spring there will not be much but garbage to place on the pile, but as summer comes on there will be garden weeds, pea vines, etc. As the pile grows, I keep building up the sides with sod or other material that will stay in place, and keep covering the succeeding layers with topsoil, or, if available, with thin layers of manure.

In the fall all the cleanup of the garden goes on the pile, squash, tomato and bean vines, the remains of the cabbages, cauliflowers, broccoli, and so forth. Then I cover the pile with sods, root side up, or manure, and leave it for nature to take its course, not touching it again until a year from the following spring. By that time, the pile is two years old, having taken six months to build and eighteen months to cure.

Now, as soon as the frost permits, I strip the old pile, placing all the uncomposted material on the new one. The balance, which is the most of it, is a crumbly, nearly odorless material which is practically one hundred per cent humus. This is the material I will use for the current season of planting.

This method of compost building should be known as "the lazy man's method," for it involves no turning over of piles, no watering, and no treating with activators, but it seems to work. It produces for me each year about three-quarters of a cord of fine compost, which is a necessary supplement to the four cords of manure that I need for my garden each year. The special uses to which I put it I will speak of later in a subsequent chapter.

The drawbacks of this system are several. First, it requires two piles, as well as space for a third, with the resultant loss of an extra year in getting started. Furthermore, it will not handle materials which are hard to decompose, such as dried leaves and cornstalks, unless they are chopped up. Nor is it completely foolproof for, unless care is exercised in building the piles, it will be found that excessively thick layers of garbage will not get enough oxygen; as a result, the decomposition may produce a black, slimy mess instead of crumbly compost. This unwanted result can be avoided, however, if care is taken to keep the layers thin and sprinkled with loose topsoil. If spots of this mucked-up compost are found in the pile, sort the material out and toss it into the pile which is currently being built.

When gardening operations are being started for the first time, there will be no compost to work with, and for the first season other materials will have to be used. I recommend that you search for well-rotted stable manure, which is the best substitute for compost. If this cannot be obtained, there are various organic products advertised by the seedsmen and garden supply houses which are suitable. A recent addition to these commercial forms of organic fertilizers is the use of treated sludge from municipal sewage disposal plants. According to Jerome Goldstein "The sale of organic soil conditioners is a big business . . . according to the United States Department of Agriculture, more than 25,000 tons of compost are already being sold, along with some 340,000 tons of dried manures and 120,000 tons of sewage sludge. (*Garbage As You Like It*, Rodale Press Inc.)

During this first year the compost pile will be in the process of being

CONVENTIONAL COMPOST MAKING

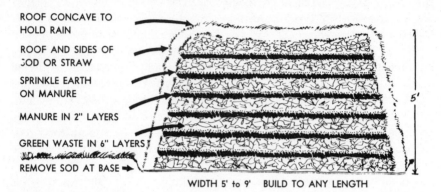

ROOF CONCAVE TO HOLD RAIN

ROOF AND SIDES OF SOD OR STRAW

SPRINKLE EARTH ON MANURE

MANURE IN 2" LAYERS

GREEN WASTE IN 6" LAYERS

REMOVE SOD AT BASE →

5'

WIDTH 5' to 9'   BUILD TO ANY LENGTH

Speed-up compost piles depend on assembling a quantity of green waste matter and a supply of manure, then building the pile in one operation. Most authorities suggest the addition of lime and/or prepared bacterial activators to assist decay. Weekly wetting down and forking over at intervals of several weeks will create compost in a single season. (A group of organic experimenters once composted a dead calf in 4 days.)

built, and the type of pile that I recommend will not produce compost in time for use in next year's garden either. Provision must be made, therefore, to supply this lack. There may be several solutions to this problem, one of which would be to build two piles of compost at once, one of them as I have described, the other built of material which you deliberately go out and search for — leaves, spoiled hay, lawn clippings, and any organic refuse that you can find. This pile will have to be given the quick composting treatment, which involves adding special solutions, wetting, mixing, and turning.

For methods of quick composting, I recommend that the reader refer to other authorities. A standard reference which covers every aspect of gardening, and which has an excellent section on composting, is the *Garden Encyclopedia* edited by E. L. D. Seymour and published by Wise and Company. The biodynamic method of composting is described in *Biodynamic Farming and Gardening* by Ehrenfried Pfeiffer and published by the Anthroposophic Press, New York. For the Indore Process see Sir Albert Howard's *An Agricultural Testament* published by the Oxford University Press.

A more recent encyclopedia than the Seymour work is one published by Rodale Press which specifically treats organic gardens, *The Encyclopedia of Organic Gardening*. In fact there are a number of treatises issued by Rodale Press, one is *Pay Dirt*, and another called *Compost* a book devoted exclusively to composting.

[48]

Rodale's *How to Grow Vegetables and Fruits by the Organic Method* has a chapter called *How To Make Compost*.

There are many other sources of information on this subject and adequate instructions are easy to come by. It may develop that you will prefer the quick composting method once you have tried it. As I have stated, the reason I use the method I have described is because it takes less of my time and labor; I have no objections to other methods.

Since having written the preceding pages on composting I have modified some of my practices, but as a matter of fact there is nothing I would change except to indicate that good compost can be made with even less trouble than the method outlined requires. I now have three permanent pits walled up with cement blocks, built in the manner described in the next chapter. I use all the garbage I can get, and find that there is no problem of odor or rats. We have birds and dogs galore, but if care is taken to keep the garbage covered there will not be any nuisance problem.

But before tying up the subject I would like to reiterate that unless you give them special treatment, keep dry leaves and lawn clippings out of your compost pile. I emphasize this point because of the fact that almost everyone who writes on the subject stresses the fact that they should be used. A recent article in the *Boston Globe* states the following: "Autumn leaves are a good starting material, but you can also use lawn clippings, straw and plant refuse from your garden or the kitchen. Some people use garbage, but this can be offensive in highly populated areas." Now I don't care who is responsible for this advice; it is wrong. If you are going to use autumn leaves they must be specially treated; either put through a chopper before being put in the pile, or they must be composted in a special pile of their own which will require considerable time to produce leaf mold, and that will be only at the bottom of the pile. Of all the troubles which the composter may have to face, undecomposed leaves are the worst, and this you must expect if you carelessly dump dried leaves in your compost. Green lawn clippings should be piled separately and cured before introducing them into the compost pile. Unless this is done the raw grass will heat too hot and decompose into a slimy mess which will eventually harden into a substance far removed from compost. And of course the advice about garbage is 100 percent wrong. These wastes are the ones which should, more than any other (except sewage) be returned to the soil. If properly handled the sweet smelling result will disclose its origins only by virtue of the broken egg shells in it.

Recently I wrote an article for *Vermont Autumn* on the subject of composting autumn leaves, and I include excerpts from it herewith.

"I never was a leaf-burner, nor

shall I ever be, for while burning is a simple and fragrant way of getting rid of leaves, it is not a sensible thing to do. In the woods the leaves fall and they form there on the earth beneath the trees an enriching and warming blanket without which the forests could not exist. Where the leaves are thickest they are not evil; it is mostly in villages and suburbs where shade trees are grown that there is a problem. But in these places as well as in the woods, nature's end would be best served if the leaves were allowed to lie on the ground. Since this is not always practicable and since there are gardens to be made fruitful, it makes the most sense to see that somehow the fertility inherent in the leaves be used to enrich the garden.

"But to get the goodness of the dead leaves into the garden soil is not as simple as it seems. Leaves are tough and everlasting and they tend to lie in layers which exclude both air and moisture, and thus they decompose slowly, from the bottom up. The most efficient way to turn them into compost suitable for garden use is to shred them, and to do this requires some sort of a machine. A rotary lawn-mower will shred them as they lie on the ground, but the best way is to gather them up and put them through a shredding machine. These are to be found for sale at almost any place dealing in farm tools. It might be practical for neighbors to own such a machine together. Once shredded, the leaves are to be mixed directly in with the other materials in the compost pile, preferably with some stable manure added.

"Another system is to spread them on the garden soil on a windless day, and chop them into the soil with a rotary tiller, thus the moisture and the worms can get at them bit by bit, and when time to plant comes in the spring, they will have been assimilated into the soil. If either of these methods is too much trouble one can make a bin out of chicken wire in some shady place, and just dump the leaves in there from year to year, using the leaf mold as it accumulates on the bottom of the pile.

"At any rate the place for the goodness of the leaves, of which the crop from one good sized shade tree is worth as much as $15.00 in terms of plant foods and humus, is in the soil. And so perhaps ordinances which on the face of it seem a bit silly, may be just the inspiration we need to get the goodness of the leaves back into the arms of mother earth, even if in so doing we lose the delights of 'smoke-scented draughts of autumn air.' "

One other source of organic material for the garden is so-called green manure. Green manure consists of a cover crop planted on the plot for the express purpose of plowing the green material under. These crops are generally forage or grain crops, such as rye or clover, or legumes, such as cow peas, soy beans, etc. This procedure I do not recommend for small gardens; the soil is out of use for a season, it requires the services of a big plow, and, unless the conditions are just right, the required decomposition may not take place.

So for our garden, let us stick to the use of well-rotted manure, if it can be got, and, if not, compost which we make for our own use. In any event, we will make compost as a complement to our supply of manure and for special uses which will be described in a subsequent chapter.

Step No. 1: New compost pile starting about May 1. (Last year's pile shows at left, still curing.)

Step No. 2: Pile is building. Use everything . . . left over manure, weeds, spoiled hay, and occasional layers of soil.

Step No. 3: Pile is finished, roofed with sod. This one has been curing for about six months.

One year later: Pile is stripped and ready to provide rich nutrients for this year's garden. Odorless and clean. Wonderful stuff!

CHAPTER FIVE

Tools for the Garden

FOR THE PREPARATION of the soil and the cultivation of the garden it is necessary to have certain tools and implements. To start, the garden soil must be turned over, and this must be done by hand or by the use of a plow. Small gardens can be handled well enough by hand, and the decision on this point will have to be made by the gardener himself. If it is to be done by hand, I recommend the use of a short-handled spading fork, with heavy tines of well-tempered tool steel. Cheap, flimsy forks will soon be prized out of shape so that the tines are not in line, and then the usefulness of the tool is ended. For larger gardens, it will be necessary to use a plow; this may be drawn by a horse or tractor, or it may be self-propelled.

Unless the horse and plow, together with a plowman, are available, the best thing to use is a self-propelled, gasoline-driven unit. The heavier tractors will compact the soil too much, and the big plowshares are apt to cut too deep. Of the self-propelled implements there are two types; one, not properly a plow, prepares the soil by means of a set of revolving tines or teeth.

Of this machine there are two distinct types, one is propelled by the revolving of its cultivator tines, which rotate comparatively slowly. If used for the chopping up of greensward or loosening hard soil, this type is at a disadvantage, for considerable effort is required on the part of the operator to hold it in place while the tines chop away in one spot. The other type has a set of rapidly revolving tines which operate separately from the propulsion of the machine. This type will do a better job on sods and hard soils, but in fine and soft soils it suffers from the very power and speed of the revolving tines. These may so fluff up the soil that it must be allowed to settle before a suitable seedbed can be obtained. Then too, if the soil is rich and full of earthworms; look out! The process of pulverizing makes no exceptions for the poor *lumbricus vulgaris*. Again, as is often the case in rich garden soils, the *bête noire* of the gardener is witch grass, also known as quack grass, and this pest propagates from underground stems of which each tiny piece will put out roots and start a new plant. For these reasons, in that I have rich and friable soil to work with, I prefer the slow revolving type. It may require more use of muscle, but it will do all required and this without the drawbacks I have mentioned. These rotary tillers have an important part to play in preparing the soil for spring planting, and in the fall they will chop the manure or compost which has been spread, into the soil, where over the winter months, nature will serve to incorporate the organic material into the very soil itself. But I should not like to have to get along without the small, walk-around-self-propelled plow, with conventional moldboard and share. This type of machine turns the soil under, so that which was on top becomes buried, and when coarse organic materials or weeds cover the surface of the garden, this turnover solves the first problem of getting a good seedbed, which is of prime importance. My practice is to chop the surface material into the soil in the fall and then to turn it under in the spring.

In old gardens, or for that matter in any garden after the first year or two, it will be necessary to go around the edges of the garden plot by hand before the ground is plowed. If there is witch grass or other root-propa-gating weed in the paths or around the edges, it is well to set a line along the edge of the garden, cut down through the roots to the depth of six or eight inches with a sharp-edged tool, then spade all along the edge for a distance of two or three feet, throwing away all the weed roots in the process. This operation will require the use of a spading fork, which has already been described, and in addition, a cutting tool and a garden line. For a cutting tool, I prefer a short-handled spade with a flat blade and a sharp edge. This tool will have many other uses and is a must in our list.

Next, and likewise an indispensable tool, is the garden line. For this I recommend a one-eighth inch nylon rope. This material is sold by the pound, and seventy-five or one hundred feet should be purchased. For stakes, substantial metal ones are best, perhaps a pair of metal rods gleaned from the junk pile, or a pair of three-foot iron stakes, one-half inch square, which can be obtained from the local blacksmith or metal shop. Otherwise wooden stakes will have to serve the purpose. These can be set firmly in the ground and the line drawn tightly without sag or curve between them. The garden line is in constant use all through the season, and it pays to have two good ones.

After the garden has been plowed it must be prepared for receiving the seeds and plants. The preliminary operation is to go over the ground

SET HOSE IN VISE SO
THAT JAWS GRIP ALONG
DOTTED LINE — SCORE
WITH COLD CHISEL—
BREAK WITH HAMMER

OGDEN HOE   SPADING FORK — SPADE

POTATO HOOK   RAKES, BROAD & NARROW   BROAD HOE

with a potato hook, sometimes called a crile. This tool will also have constant use and it is imperative to have a good one, preferably one with slightly flattened tines. Next come the rakes, and we will need at least two of them. Both can be bought at any hardware store, but one will have to be altered to meet our specifications, for I have been unable to find any manufacturer which makes one as narrow as is required. For the first, any sturdy steel rake will do, as wide as possible, with eighteen or more teeth. The other is a lighter one, as narrow as possible; it will probably have to be cut down,

for we want one not over eight or ten inches wide. The uses to which this special rake will be put will be discussed later.

You will also need two hoes, one broad-bladed for hilling up and general cultivation, and the other, preferably with sharp corners, for working close up to and in between plants in the row. This latter can be a "Warren hoe" which, being more or less heart-shaped, has but one point; or it can be one which I have described in an earlier book as the "Ogden hoe," named for my Uncle Will who devised the first one I ever saw for use in his own garden. He

[57]

would take an old, well-worn hoe, set the jaws of a vise along a line drawn from one corner of the blade to the point where the tang joins the blade, then with a cold chisel cut off the material projecting beyond the jaws of the vise. The operation was then repeated on the other side of the blade, so that a triangular blade with a sharp point at each lower corner resulted. Actually, I do not use a hoe very much in the working of my garden; its principal use is in the planting of corn and peas and in hilling up certain vegetables after the plants have achieved several inches of growth.

Of miscellaneous tools there will have to be a manure fork, either short or long handled, for handling manure and compost and cleaning up the piles of vines and garden refuse; also a long-handled, round-pointed shovel for which we will find many uses, such as trenching, digging holes, banking up celery, etc.

Since you run an organic garden there will be no need for any apparatus for dusting or spraying, but if fainthearted concerning the control of pests and disease and the use of rotenone or pyrethrum is indicated, a dust blower is all that is

MANURE FORK   SHOVELS

LINE

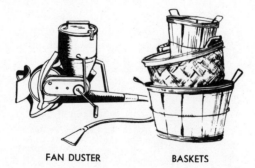

FAN DUSTER   BASKETS

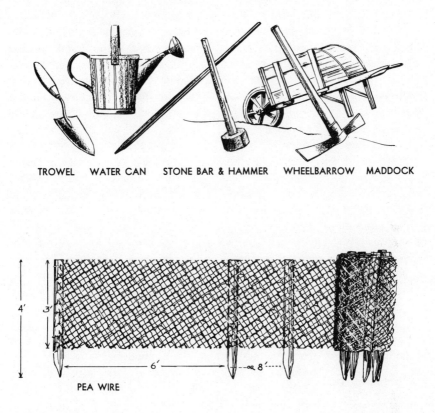

TROWEL   WATER CAN   STONE BAR & HAMMER   WHEELBARROW   MADDOCK

PEA WIRE

necessary. Of these, the best type is that which hangs from a shoulder strap and which, in addition to a receptacle for the dust, has a fan-type blower propelled by a crank. This machine comes with several types of nozzles which can be fitted to the end of a flexible tube. It does not require frequent recharging, and the design makes it simple to direct the dust against the under sides of the plant leaves.

It will be advisable to have on hand a varied assortment of baskets; experience will indicate the sizes and shapes which come in handy when transplanting, gathering stones, or harvesting vegetables. For transplanting and small work, an ordinary garden trowel is necessary, as is also a watering can. I also keep on hand a hardwood stick about one by one-quarter inches, and five feet long, on which are marked with knife cuts all the different widths of row spacings which I use. I also have a contraption of my own design consisting of two pieces of one-quarter by one inch iron attached to a wooden handle. The irons are bent so that when dragged behind me the ends make a double set of marks thirty inches apart in the soil of the garden. I use it for marking out the checkerboard used in planting the hills of corn.

If you plan to grow peas, it is well to have on hand the material for the vines to grow on, and it is not necessary to go through the procedure of gathering this material together each year. If you have two garden plots as suggested, it would be well to have them both the same width, so that wire prepared for peas in one plot can be used in the other as well. Get enough galvanized chicken wire so that it can be cut to the proper length for as many rows as you plan to grow of peas. Then at the lumber yard, procure a supply of two by two scantlings. If you follow my advice in later chapters, three-foot wire will be the proper width, so the scantlings should be cut into four-foot lengths, and one end of each sharpened with a hatchet. The lengths of chicken wire should then be stapled to the stakes, one edge flush with the square ends. The length of the rows will determine how far apart the stakes will be placed, which should not be over eight feet, and they do not need to be closer together than six feet. The stakes on each strip should then be painted with some decay-preventing preparation and rolled up and stored for future use.

These pea supports can be used year after year. Roll them out along the row before planting; then, starting at one end, set each stake firmly, using a maul to drive them in and stretching the wire tight as you go. When the peas are gone, clean out the vines, then pull out one stake after the other, laying the wire flat as you go, finally rolling it up and storing it for use again next year.

No discussion of garden tools and impedimenta would be complete without mention of the wheel hoe or cultivator. I note it merely to dismiss it as it is an expensive piece of gear and one for which we have no real need. Furthermore, for the home-sized vegetable garden, there is no need to invest in a push-around type of seed drill. Actually, the best seeding can be done by hand, as can the best and quickest row cultivation, and with less effort at that.

I have spoken of peas, and it might be well to prepare in advance for two other members of our garden family, beans and tomatoes. If you plan to grow either pole limas, Romano or Kentucky Wonder snap beans, it pays to have the poles on hand before planting time. Pole limas can be grown only in climates having a long growing season, but Kentucky Wonders and Romas can be grown nearly anywhere. Both beans are worth growing, so, depending on your location and climate, it would be well to plan for one or both. The number of poles will depend on your garden plan, which we

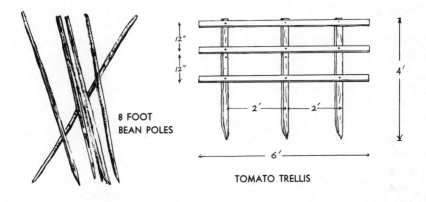

8 FOOT
BEAN POLES

TOMATO TRELLIS

have already discussed. They should be at least eight feet tall and sharpened with a hatchet at the butt end. Cedar poles are the best, for they are the most resistant to rot. They can often be purchased at a lumber yard or farm supply depot, but if you live in the country, you can get them out yourself, using whatever kind may be available. They should be straight and of even length, not over two and a half inches through at the butt, and not much less than one inch at the tip. I use balsam fir poles, which I have discovered will last about six years before the butts rot off.

For tomatoes I use sectional trellises which can be pulled up, stored, and used year after year provided they are taken care of. I

have grown tomatoes without supports, but experience has proved that it pays to spend the extra effort required to keep the fruit off the ground. The easiest method is to use a trellis, which should be in sections, each one about six feet long, consisting of three stakes made of two by two scantlings four feet long, sharpened at one end, and three parallel cross slats, each six feet long, and made of one inch by two inch lattice strips. Each middle stake should be in the center with the two end stakes twelve inches in from the end of the slats. The top slat should be firmly secured with nails or screws flush with the square end of the stakes; the next slat should be one foot below the first, and the bottom slat one foot below the middle one.

The whole should then be painted with some wood preservative and stored away ready for use when the time comes to set out the tomato plants. Each section will take care of three or more plants, and the garden plan will indicate the number of sections to have on hand.

Before leaving the subject of tools and going on to a description and the uses of a cold frame, let us gather up some tools, which, while not absolutely necessary perhaps, are useful things to have around. A steel crowbar, not too heavy, comes in mighty handy in setting bean poles, and with it we can drive holes in our compost pile if we think it necessary to do so. It is also useful in removing large rocks from the garden. Likewise, a ten- or twelve-pound stone hammer or maul will come in handy in setting stakes or breaking up rocks too big to move in one piece.

I had written in the earlier edition that a wheelbarrow has so many garden uses that it could be con-

Looseleaf lettuce ready for early spring use having been transplanted from garden to cold frame the previous fall.

[62]

Swiss chard, lettuce and onions ready for early spring use having been transplanted from garden to cold frame the previous fall.

sidered as indispensable, and this is most certainly the case, except that now, something really new has been added to the line of vehicles used for transportation of garden items. The two wheeled, ball-bearing square garden cart, of which there are two varieties made in Vermont and nationally advertised, is so superior for garden use, that for these purposes it has made the wheelbarrow obsolete. The wooden wheelbarrows which I described in the earlier book which were hard to locate then, are practically impossible to find now.

So the two-wheeled garden cart is a happy solution to the problem, and it can be used for so many household uses as well, so that it really becomes a required item.

I have changed my mind concerning the cold frame as well. Previously I described it as important but not essential, and while that must still stand, I think now that every garden really warrants the inclusion of a cold frame even if it is a very small one. It will not be necessary to have one of the size described here, but even if it should be so tiny as to be made up of but a single storm sash, I think it would be worth having.

The expense of buying tomato, cabbage, pepper, cauliflower, etc., plants may not amount to much in a tiny garden, but if many plants are used, especially of tomatoes, the saving in growing your own is considerable. At one time I used to purchase my plants from the greenhouse, but I have not done so for years, and I wonder that it took me so long to discover the advantages of growing my own. First of all you have a complete choice as far as varieties are concerned, and above all, you have a reserve of plants with which to replace the depredations of the cut worms, or losses from any other cause.

A cold frame is an important, though not essential, adjunct to every home garden. Its uses are several: it can be used for hardening off plants which have been started in flats in the house for transplanting early in the spring; it can be used for the early planting of frost-tender vegetables; and it can be used as a bed for transplants in the fall, so as to carry them through the winter for table use early in the spring. I put my cold frame to the first and last of these uses. The last thing in the fall I fill my cold frame with transplants from the garden — lettuce (looseleaf), endive, parsley, and Swiss chard. All of these will survive the winter in the cold frame and so we have greens and salads for table use before the snow is all gone in the spring.

If, for one reason or another I fail to set plants out in the fall cold frame, I fine the soil and broadcast lettuce seeds. If the timing is right germination will take place the first thing in the spring, and there will be a plentiful supply of lettuce plants ready for transplanting as soon as the soil is in proper shape.

I have had several cold frames and have finally developed a design which is not only easy and economical to build but is permanent as well. Cold frame sash can be purchased at any building supply depot. They come three feet wide by six feet long, so the size of your cold frame, depending on your needs, will be in multiples of these dimensions; that is, either three by six, six by six, nine by six, or like mine, twelve by six. The frame itself is to be built of cement or cinder blocks, the dimensions of which are eight by eight by sixteen inches, and for our purpose can be laid up dry; that is, without the use of mortar. Their weight will hold them in place, the laying of them will present no problem, nor will the services of a professional be required. The front wall will be two courses high and the back wall will be four courses high, so that the pitch of the sash from back to front will be sixteen inches, which is ample to provide for runoff. The length of the front and back walls will depend on the number of

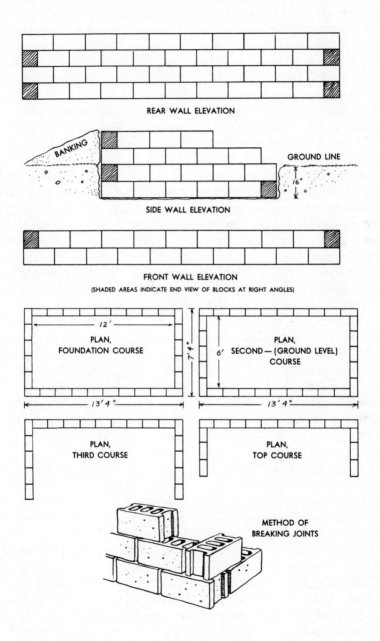

REAR WALL ELEVATION

BANKING

GROUND LINE

16"

SIDE WALL ELEVATION

FRONT WALL ELEVATION

(SHADED AREAS INDICATE END VIEW OF BLOCKS AT RIGHT ANGLES)

12'

PLAN,
FOUNDATION COURSE

7' 4"

6'

PLAN,
SECOND — (GROUND LEVEL)
COURSE

13' 4"

13' 4"

PLAN,
THIRD COURSE

PLAN,
TOP COURSE

METHOD OF
BREAKING JOINTS

sash used. The two end walls will be the same length in any case and will require five cement blocks for each of the three bottom courses and three for the top course, making a total of eighteen blocks for each wall. Now, depending upon the length of the bed (I do not recommend anything less than two sash), the number of blocks needed will be as follows: For a four-sash bed, two courses in front of ten blocks each, and four courses in back of ten blocks each, making a total of sixty blocks. A two-sash bed will take half of this, or thirty blocks, so we can figure a total of ninety-six blocks for the four-sash bed and sixty-six blocks for the two-sash bed. Additional material in the form of two by four scantlings will be required as shown on the diagram. Having the materials now on hand, it is time to start construction.

First, excavate a hole sixteen inches deep, the size depending on the number of sash used, making sure that it is large enough to allow for the eight-inch-thick cement block wall all around. Now lay up the walls, checking to be sure that the first course is level, that each block sits firm and true, and, most important, that the inside dimensions all around compare exactly with the dimensions of the sash to be used. Now fill all the holes in the blocks with earth and tamp firmly before starting the next course. At each corner one will lap the other butt, breaking joints on the second course, so that the second laps where the first butted, etc. A study of the accompanying drawing will clarify this procedure. Continue in this manner until the walls are laid. With the earth from the excavation, bank up around the back and side walls until the banking is flush with the top of the walls. The front wall, of course, will be flush with the ground, or, better, an inch or two above the surface. Fill the bottom of the pit with six inches of well-rotted manure on top of which will be placed good, rich topsoil which has been mixed with compost or composted manure. Now we are ready for the frame which will support the sash; it is simple in construction and is made of two by fours, consisting of front, back, and side sills with cross supports depending upon the size of the bed. Again, details of this construction can be obtained from the drawing. The sash, which should be hinged to the back sill, can be raised and lowered at will, with notched props to hold the sash open as needed.

This type of cold frame, a permanent and valuable addition to any home garden, can be built at a minimum of expense by using your own labor, and within a few years it should pay for itself. As a substitute for sash, use rigid sheets of fiberglas.

There are other tools and devices than these which we have mentioned and no doubt some of them will find their way into your collection before many years pass by. There will be

SASH FRAME, CONSTRUCTION AND SUPPORT

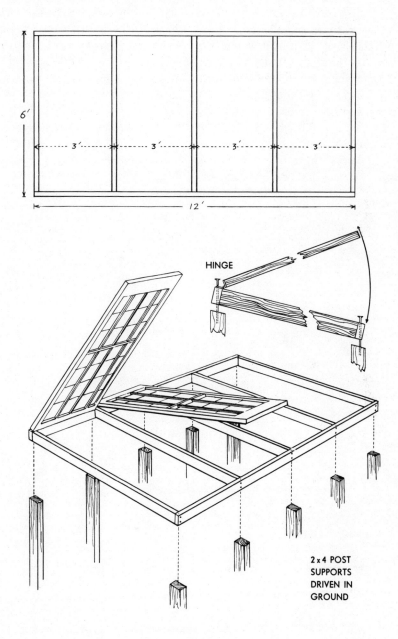

HINGE

2 x 4 POST
SUPPORTS
DRIVEN IN
GROUND

duplicates of some, for if you have help in your garden work, you may want two potato hooks in service at the same time, and the same goes for other tools as well. I would find it hard at times to get along without a pickax or a mattock (sometimes called a grub hoe), but these are not indispensable, nor are the countless new gimmicks which appear from year to year in the hardware stores and then disappear, although among them we may find one that is useful to us. As a further refinement for those who have the time and interest, a hotbed can produce results not otherwise obtainable, and an interest in this side of gardening logically leads to a greenhouse, which opens up a whole new and fascinating field of gardening. In describing tools I have limited myself to those which I have found indispensable in the working of my own garden, but I acknowledge that there is room for great latitude in the selection of garden tools.

CHAPTER SIX

Preparing the Seedbed

NOW WE ARE READY to get down to the actual business of making our garden, and whether it be an old or a new one which we are just starting, the work must begin in the fall. If it is a new garden, the plot will have been selected, cleared of long grass, weeds, brush, and rocks. We are then ready for the first step.

If it turns out that the most logical location lacks an adequate amount of good topsoil, this first step may consist of a trenching operation. This laborious bit of business, which has been described in detail in Chapter Two (page 14), should be undertaken only if the soil is hard-packed and impervious so that ordinary spading goes little below the surface (or the subsoil plow may be used, as we have previously indicated). Once the soil is in proper organic condition, we will not have to concern ourselves with this problem, for worms and burrowing insects and animals will do the job for us.

In another instance, if the garden

is to be a small one and the labor involved is not too great, it would be well to remove the sod completely. This can be done by cutting the sod into strips, using a sharp-edged spade. These strips, which can then be peeled off in rolls, should be set aside for the building of the compost pile. Sod makes excellent material for composting so it will not be lost to us.

If it is not practical to remove this top layer of grass and roots, the sod will have to be plowed under, in which case there may be a scarcity of good loose earth for the first season. On the other hand, if the garden site is free of old matted sod and has a reasonable coverage of topsoil, the first step will be to plow the site, or better yet, to turn it over by deep spading, for when the operation is done by hand all the lumps can be broken up and the rocks removed as they are come upon. Sir Albert Howard claims that hand-spaded lands give better yields than those readied by mechanical means, though he does admit that this fact, if it be a fact, has never been adequately explained.

The next step is to get the manure or compost on the land. In my garden I save my compost for use in hills and trenches, using only manure for broadcasting. If compost is used in the rows and hills, it is not necessary to use as heavy a charge of manure as it would be otherwise.

In the case of a first-year garden there will be no compost on hand, either for broadcasting in the fall or for use in rows and hills in the spring, so it will be necessary to procure from some source or other a sufficient supply of manure for this important first covering in the fall. Speaking of the amount of manure to be used, W. W. Rawson recommends its application at the rate of twenty cords to the acre for use on market gardens where two crops per season are to be grown. This is just about the degree of concentration that I like to use on a garden, and it figures out roughly to be one and one-third cord for each plot fifty by fifty feet in area. A cord is a pile eight feet long, four feet high, and four feet wide.

Manure spread thus in the fall, evenly, with all lumps broken up, forms a blanket which prevents the frost from penetrating as deeply as otherwise it might, and it tends to keep the earthworms closer to the surface of the ground. Speaking of worms reminds me of another strong objection which I have to the use of rotary types of tillers for the preparation of garden soils. In the process the whirling blades, which revolve at high speed and with tremendous power, literally massacre all the worms in the important top layer of garden soil. Earthworms are of untold value to the gardener because of their work in aerating, draining, and enriching the soil, and their wholesale destruction is surely to be avoided.

Do not forget this fall application of manure will be plowed or spaded under in the spring; it is imperative, therefore, that it be spread evenly. If the application is too thick in spots, or if there are unincorporated masses of bedding, hay or straw, therein, it will not turn under properly, and the result will be difficulty in preparing a proper seed bed when the time comes for doing that.

In summary then, the fall work consists of clearing the land and perhaps peeling off the sod, if the site is a new one; trenching, or subsoil plowing, if the lack of good topsoil warrants the effort; and clearing off all remains of the season's crops if the garden is an old one. After these things have been done the soil is turned under, either by plowing or spading, and then the whole area is covered evenly and smoothly with an adequate charge of stable manure.

With the coming of spring the real business of making the garden begins. As soon as all the frost has gone and the ground has dried out

FALL PROCEDURE, NEW GARDENS

1 CLEAR THE SITE

GRASS, WEEDS, BRUSH

ROCKS, STONES

Site selected, plots laid out. Brush and weeds removed by scythe, brush hook sickle. Tackle the rocks with stone bar, hammer, stump puller or drag chain

2 BREAK UP THE GROUND

BY TRENCHING

OR BY STRIPPING OFF SOD

OR BY HAND SPADING

The hard part. But here is key to next summer's success. Be thorough.

3 SUPPLY THE FOOD

BUY THE MATERIAL

SPREAD IT ON

The simple part. Yet search for manure often most frustrating of all. Small gardens need only one cord or less.

[71]

to a certain extent, it will be time to take the first steps. Before doing so, however, it will be well to review what we had in mind when the planting guide or plan was set up.

First of all, remember that in preparing the seedbed the ground must not be worked too far in advance of the moment when the seeds are actually planted. Thus it will be necessary to have in mind the order in which the various seeds will be sown. There will be a spread in time of as much as a month or more between the planting of the first seeds

and the last, and we do not want to work over the ground for beans, for example, at the same time that we prepare the soil for peas.

In order to get properly organized, then, it will be necessary to prepare a planting sequence. This will be the result of balancing three main factors: which portion of the garden is first to arrive at the proper condition for working, which plot will receive a certain group of vegetables so that our crop rotation will be sensible and practical, and finally, which seeds will go in first, as condi-

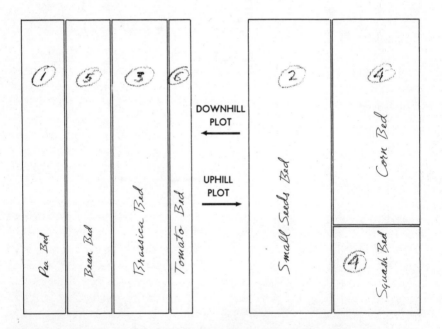

An example of planning which at planting time, might call for modification. Above, two-plot garden described on page 28. Planting priority dictated by species and season requires beds be prepared in sequence indicated above. However, such jumping around may not be practical, except if hand spaded. Again, uphill plot likely to dry first, thus peas better planted in space planned for corn.

A large yield on a relatively small garden plot is the result of careful planning and close placement.

tioned by our desires for early results and the ability of the plants to resist frost.

In setting up the plan, we anticipated the advisability of crop rotation, but now at this moment we must consider the condition of the garden, which may be ready to work in one spot but is far from ready in another. In general, the order in which we want to get our seeds in is peas first, and simultaneously with them, or slightly later, most of the small seeds, followed by our Brassica group of transplants, along with onion plants, if we are to use them; then, when danger of frost is over, corn, beans, tomatoes, squash, cucumbers, and late plantings of small seeds. In view of the condition of the soil and the need for adhering more

or less strictly to the sequence of planting as described, it may now be necessary to rearrange the garden plan. We will start with that portion of the garden which is ready to be worked, even if in our plan we had this specific spot reserved for corn, which will not be planted for perhaps another month.

In deciding when and if the soil is ready to be worked, do not attempt to work heavy, clayey soils while wet, for they will cake up in hard lumps and will not permit us to make a proper seedbed. Sandy loams or, as a matter of fact, any type of soil which is rich in humus and in good tilth can be worked as soon as the frost is out and the sogginess gone.

So, no matter what the plot plan

SOIL PREPARATION, SPRING PROCEDURE

a. Spade or plow under manure spread previous fall.

b. Potato hook next. Work at right angles to plow furrows.

c. Look to the tools, trellises and supports. Ready the compost.

d. Final smoothing with hook and rake at moment of planting. (see text)

says, start working the plot which is ready first, at the side which is least moist. Spade or plow under the manure. This must be done carefully so that all of it is turned under. Throw off the rocks and stones. The new surface should be smooth, friable dirt. The plowing or spading should be done crosswise to the slope of the garden. The conventional procedure is to follow plowing by harrowing. Harrow attachments can be had for walk-around tractors, but in my experience, once the soil is in good condition it will not be necessary to harrow after plowing. The soil should come off the moldboard smoothly and evenly, breaking up as it falls against the other furrow. Working over it with the harrow would only serve to compact it.

In lieu of harrowing, I go over the whole plot carefully and thoroughly with a potato hook. Starting at the upper edge of the plowed piece and working backwards so that all footsteps are worked over in the process, I move at right angles to, or across, the furrows left by the plow, smoothing out the inequalities, breaking up lumps, and removing clods, stones or what-have-you as I progress.

Now all is ready for the preparation of the seedbed, but before we start we have a few chores to perform in preparation for the actual seeding. First we check our garden tools and ready them for use; we get the pea wire out of storage and prepare the compost pile, for we will have to use compost in the planting of peas. After the compost cycle is in full swing, there will be a new pile ready for use which will have to be prepared, or at least which will require some attention before we can place it in the trenches for the peas. As soon as the frost is out of the outside layer, which may consist of sods or a mass of weeds and roots, this shell should be peeled off and placed in the new compost pile. The core of the pile is then ready to be shoveled into a wheelbarrow for transportation to the site of the planting. In turning over this material all uncomposted stems, bits of sticks or stones, etc., should be thrown out. This operation requires some care but it is less laborious than screening, and if the pile is well composted it is all that is necessary. The material you now have should be a dark-colored, sweet-smelling substance, light and crumbly and plentifully sprinkled with worms.

Having assembled our materials, we proceed to the final preparation of the bed, which consists of going over it with the iron rake, fining and smoothing the soil to the last degree. The soil will now be rid of lumps and stones, and will be loose and friable for the free movement of air and water, yet will be of sufficient density so that there will be close contact with the seeds after tamping, to the end that there will be full and rapid germination. This process will be gone into in detail directly, but first I will mention the exceptions to the general procedure. In the case of peas, it will not be necessary to do any more than go over the soil with the potato hook as described. Now the wire can be set and the trenches dug and filled; then, after the planting, the space between the planted rows will be gone over carefully, first with the potato hook, if the soil has been compacted in the process of planting, and then with the rake so as to prepare a fine seedbed for the planting of small seed between the rows. In the case of transplants, there is no necessity for fining with the rake, nor will the corn or bean patches require more than the preliminary going over with the potato hook.

For the small seeds, however, it will pay to prepare as fine a seedbed as possible. To achieve this, I go over a strip about three feet wide with the potato hook and in the opposite direction to the first. This takes place immediately preceding the actual planting to assure that the dried out surface is well mixed with the moist soil beneath so that the seed will not be dropped on dry earth. Now I walk around the garden to the head of the strip just worked over and, working in the same direction and moving backwards as always, with the rake fine a strip of sufficient width to plant one row or a little more. In all of these preparations there is one basic principle which must not be forgotten, and that is to avoid as much as possible walking on the garden during any stage of the preparation. Furthermore, for looks, if for no other good reason, I always govern my movements in the garden so that no footprints or tracks are left between the rows or anywhere else; I work my way out so that there is no disfiguration of the smooth surface of the garden.

The ideal way to handle the preparation of the seedbed is to spade or have plowed only that portion of the garden which will be planted immediately. Having two or more plots in our garden, and in accordance with the garden plan, it will probably be necessary to have a part of each plot plowed before any one of them is entirely seeded. In many cases this may be an impracticality. If the garden is a small one and you spade up the earth yourself, seeding can be accomplished in this manner; but if the plowing has to be done by someone else, the chances are that the only sensible and economical way is to have the whole garden done at one time. When this is the case, the secondary working over with the potato hook becomes more important than ever, for it is by this means that plowed land, which has lain for some time exposed to sun and wind and the sprouting of weeds, is loosened up and aerated, a prerequisite for a proper seedbed.

I hesitate to make explanations to the point that what in reality is a simple and commonsense procedure seems to become complicated and difficult. It is possible, of course, to wait until all danger of frost is over, then plow the entire garden plot, harrow it, and, in a great burst of energy, get all the planting done and over with in one fell swoop. This might seem to be the simplest way, and I know that in many instances it is the way taken. But even though it seems to be simple, such a procedure results in complications later on, for all the weeding, thinning, and cultivating operations would have to be done at once as well, and inasmuch as these later operations

SUCCESSIVE PREPARATION

Lower bed turned over, pea wire up and peas in. Uphill beds waiting for the plow.

are more laborious and exacting than the planting, the result would be that they would get done only in part, and not properly at that. In the end, therefore, what appears to be the simple way is in reality no way at all. In addition to this, it would mean that many crops, which could be early ones, turn out to be late ones, and in some instances the late plantings would result in no crop at all. For instance, the seed of lettuce will not germinate if the temperature of the soil is over sixty degrees. So, while we will endeavor to keep early plantings together as best we can in order to have a proper sequence of plantings and to arrange for a suitable rotation of crops, it will probably be necessary to start planting one part of the garden and then, before that is finished, move to some other part and start planting there. I have four sections in my garden. I will start in one of them with peas and certain plantings of small-seeded, early vegetables such as lettuce, spinach, radishes, and early beets and carrots. But I want to reserve space in this plot for beans, or squash, perhaps, so before that plot is finished, I will have moved on to another where I will want to get out my transplants as soon as the

[77]

TWO PLOT GARDEN PLAN DESCRIBED ON PAGE 28

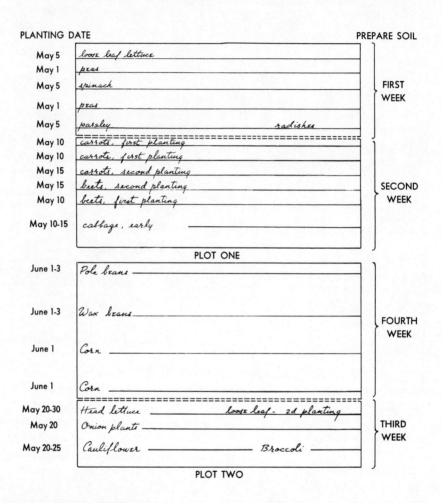

PLANTING DATE   PREPARE SOIL

May 5	*loose leaf lettuce*	
May 1	*peas*	
May 5	*spinach*	FIRST WEEK
May 1	*peas*	
May 5	*parsley* *radishes*	
May 10	*carrots, first planting*	
May 10	*carrots, first planting*	
May 15	*carrots, second planting*	
May 15	*beets, second planting*	SECOND WEEK
May 10	*beets, first planting*	
May 10-15	*cabbage, early*	

PLOT ONE

June 1-3	*Pole beans*	
June 1-3	*Wax beans*	FOURTH WEEK
June 1	*Corn*	
June 1	*Corn*	
May 20-30	*Head lettuce* *loose leaf - 2d planting*	
May 20	*Onion plants*	THIRD WEEK
May 20-25	*Cauliflower* *Broccoli*	

PLOT TWO

Planting priority indicates plots be prepared in progressive sequences shown at right. For practicality, perhaps plow entire upper plot one weekend, lower plot two weeks later. Plot No. 1 best situated where soil earliest in condition.

weather permits. Before that plot is finished I will move on to the third wherein I will plant later seedings of small seeds. Then I will go to the piece reserved for corn and plant that before I go back and finish out the first section with plantings of beans and squash. Thus, in actual practice, the plot which was started first may be the last to be finished.

In later sections covering planting and in the discussion of the individual vegetables, many aspects of the time to plant and where to plant, if there is a choice, will be gone into so that the reader who combines this information with the simple requirements of crop rotation will find the operation not nearly as complicated as the clumsy use of words may have made it seem. At any rate, I am certain that, after a very little experience, the preparation of a planting chart will present no real problem, and that, after a few years, even a chart will no longer be a necessity in the planning and growing of a good garden.

To speak again briefly of the problem of rotating crops, a subject discussed in Chapter Three, it is most important that the members of the Brassica group (cabbage, cauliflower, Brussels sprouts, and broccoli) be planted in a different plot each year. These plants are vulnerable to certain soil-borne diseases such as "damping off" and "club root" which can be much more easily controlled if the plantings do not succeed themselves in the same soil, even for one year.

If the garden contains but two plots, a sensible rotation would be to have the peas, small seeds, and, if there is room, cucumbers and squash in one plot, with the cabbage family, corn, beans, and onion plants in the other. Thus peas, radishes, looseleaf lettuce, and spinach will be early plantings in one plot, while early cabbage and onion plants will be early plantings in the other.

All in all, the preparation of the seedbed will have to be tied in with the planting, but the techniques remain as I have described them. If the seedbed is properly prepared, if the seeds are of good stock and of tested germination, when properly planted the result will be even rows of healthy seedlings. Important as is the quality of the seed and the proper placing of it in the ground, even more important is the preparation of the seedbed. When improperly done, it is the cause of most garden failures.

We cannot leave this discussion of soil preparation without consideration of the end of the harvest and the thought which must be given to the coming season. During the late summer and fall there will be work to do which is in the nature of preparing the soil for next year's garden. Pea vines and wire will have been taken out earlier, as soon, in fact, as the bearing is finished. But at the end of the season, cucumber, squash, tomatoes, and beans will still be in the ground and will have to be pulled out and placed on the compost pile. The bean poles and tomato trellises will have to be removed and placed in covered storage. Cornstalks will either have to be pulled or cut off close to the ground. I generally cut mine and then, in the spring after plowing, hook out such stubble as has not rotted away, gathering it up and placing it on the compost pile. The stalks are difficult to compost; if means are available

FALL PROCEDURE, ESTABLISHED GARDENS

THE CLEAN UP

Vines and plants pulled for the compost pile.

Trellises and poles stored for the winter.

Weeds scuffed, garden dressed for plowing.

LATE SEASON CROPS

Endive, parsley, lettuce in cold frame.

Root crops dug for the root cellar.

Left-overs to the compost. Roof up pile.

SOIL PREPARATION

As in new gardens, soil is turned.

Spread on manure. Let it sit.

to have them chopped up, composting presents no problem. Lacking such means, however, I give mine to a neighbor to use as fodder for his cattle or they can be piled separately and allowed to lie for several years. In addition to these there will be still in the ground broccoli and Brussels sprouts stalks which must be pulled and lugged off to the compost pile.

On parts of the garden there will be a good crop of weeds which I scuff off with a sharp hoe, rake up into piles, and place on the compost pile along with the rest of the garden refuse. Tender weeds such as chickweed and purslane I generally leave to be plowed under; being easily decomposed, they make good green manure. Root crops which have not been used, such as beets, carrots, turnips, parsnips, and salsify, must be dug and stored in the root cellar. If any endive and looseleaf lettuce remain from the late plantings, I transplant a goodly supply into the cold frame for use early next spring.

Parsnips and salsify can be left in the ground over winter, thus actually improving their flavor. To leave them there, however, would interfere with the fall plowing, which, to my thinking, is more important than their spring use. It is also wise to transplant some parsley to the cold frame and to pot some of it to keep in the house for winter use.

The garden is now cleared of its growth, the gear and tools are all stored away, and the beds, plowed and snug under their blankets of manure, are ready for the winter. We must give thought to the compost pile as well. It will have received a copious charge of garden refuse, which must be tramped down and compacted and the whole covered. If sod is available, it makes an ideal topping; if not, a good thick layer of manure will serve the purpose. From now on until long after frost the processes of nature will have their way in the pile and it will gradually diminish in height as the settling and decomposition proceed.

The frost, the fall rains, and the sun will work on the soil, and the snow, sometimes called "the poor man's manure," will bring its charge of available nitrogen; the worms, moles, insects, and bacteria will do their work as conditions permit, so that, when the spring rolls around in due course, the soil will be found to be in better condition than it was the previous year.

We can hardly wait for the snow to melt and the frost to go so that we can start in on a new gardening year. Each year's experiences will modify our practices to some extent; we will discover that we planted too much of this, too little of that; and we will be tempted to experiment with new strains of varieties, so that each new year is a challenge and an adventure.

CHAPTER SEVEN

Planting the Garden

WITH THE SOIL in proper condition to receive the seed, the time has come to start planting. In planning for this, the lay of the land and the points of the compass should be taken into consideration. In the first place, the rows should be crossways to the pitch of the land so that they form miniature dams which will tend to slow up the runoff in the event of heavy summer showers. Placed otherwise, the spaces between the rows would tend to serve as channels for the water and the result would be gullying and loss of topsoil. If the pitch is to the south, the rows will run east and west. Perhaps the ideal would be to have them run north and south, so that one row would shade the next less; the first consideration is the more important of the two, however, and, as far as I can observe, it makes little difference whether the rows run east, west, north, or south. The direction of the rows having been decided, it will be well to have the plow furrows run in the same direction, for then the first working of the soil with the

potato hook will be crossways to the rows, and the final working will be in the direction of the rows, as it should be.

The next consideration is the spacing between the rows. In spite of what the authorities may have to say on the subject, my objective is to get the rows as closely spaced as possible. There is a tendency to get the rows widely spaced, believing that this makes for easier cultivation; they are either very far apart to permit the use of horse- or power-drawn cultivators, or slightly less wide so that a wheel hoe pushed by manpower can be used. I do not agree with either of these procedures. It is a waste of land and of productivity to use horse- or motor-drawn cultivators on anything smaller than a commercial market garden. As to the wheel hoe which is pushed around by hand, I believe I can demonstrate that its use involves extra labor and time as well as a waste of valuable garden space. In the first place, the design of the implement requires that the energy be

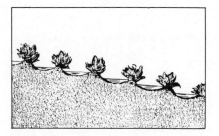

Rows planted cross-slope provide minia-ture dams, check run-off.

Rows planted down-slope result in gulley-ing, exposed roots, lost topsoil.

applied in the most inefficient manner possible. The handles are high so that all the push is from the arms and shoulders, whereas the push should come directly from the legs, as any football player knows. The result is that, in order to make progress with the machine, it is necessary to proceed by a series of jerks and pushes, which not only wastes energy but makes it difficult to control the cultivator blades with any degree of accuracy. Conse-quently, it is impossible to cultivate close to the row without the hazard of ripping out some of the plants, and after all that energy has been spent, one still has to get down on hands and knees and weed by hand in order to do a good job. If a machine could be devised which could be pulled rather than pushed, the result would be more in line with human mechanics, and to cultivate by pulling is the method I advocate.

My system is to cultivate with the potato hook, straddling the row, one foot on either side, moving back-

Plow cross-slope.

Potato hook follows up and down slope.

Raking and plant-ing of rows, cross-slope.

[83]

ward, pulling the hook as one progresses. After a little practice, one can become expert at this method of cultivation so that there is little, if any, need for hand weeding, and the speed acquired will exceed that which is possible with a wheel hoe. Using this method, I find it unnecessary to do any hand weeding until it comes time for thinning, and then I thin and hand weed the row in the same operation. Thus it is possible to place the rows of the small-seeded vegetables close together. I would like to keep this distance down to ten inches, but will settle for a maximum of twelve. If the soil is rich, as we plan to have it, there will be no lack of nutrition for the vegetables, valuable space will be saved, and most beneficial of all, as soon as the plants in the row have achieved some growth, the tops will touch from one row to the next, thus shading out the weeds and preserving life-giving moisture, which is all that could be hoped for from the more laborious and cluttery procedure of mulching.

I know that this sounds heretical, but give heed just for a moment: First of all there is a senseless wasting of productive garden space, for in the planting of small seeds it will not be possible to mulch if the rows are kept to a maximum of twelve inches apart. And how are you going to scatter small seeds such as lettuce shallowly in a fine seed-bed if you have to poke down through several inches of mulch to do it?

Besides the mess there are other drawbacks such as preventing the warm sun from striking the soil. Mulching has its many proper uses, but in the small kitchen garden it should be limited to a spreading on in the fall, with a subsequent removal or plowing under in the spring. The cult of the mulch derives from the hope that one may have a good garden without any work to speak of. No more misleading slogan could be followed; a good garden is the result of painstaking work, and plain common sense will dictate how much there must be of it. Actually if man-hours of labor are related to the total product grown, I doubt very much if it can be shown that there is any advantage to mulching between the rows. Furthermore, if they be related to product per unit of land (and that should be our goal) the result will be woefully short of that which can be accomplished by close rowing and companion cropping.

So for the planting of our small, seeded vegetables, and this includes radishes, lettuce, parsnips, salsify, turnips, rutabagas, carrots, Swiss chard, beets, spinach, and endive or chicory, I recommend a spacing of not less than ten and not over twelve inches. My procedure in planting is as follows: Starting at one end of the row to be planted, I work over a strip of soil about three feet wide, along the direction of the row, using the potato hook. When I finish the row, I walk around the garden so as not to trample on the soil, and, taking my hook with me, I return to the other end of the row, lay down the hook, and pick up the rake. With a penknife I have already cut a notch on the handle of the rake marking the distance between rows. Now I work over the same strip of soil with the rake, working along the direction of the row this time, smoothing the soil. When I reach the end of the row, using the mark on the rake handle, I set one of the stakes of the garden line, then return to the other end of the row walking around taking the rake with me and again avoiding walking across the garden. Now I carefully measure and set the other stake, drawing the line taut along the surface of the ground. I have the seed at this end of the row, and a small shallow bowl into which I pour some of the seed. Stepping out a good stride on the garden soil, I squat, setting the bowl down beside me, reaching to my left with my right hand I make a shallow furrow along the line with my finger; then, with seeds taken from the bowl between my thumb and forefinger, I sparingly scatter the seeds in the drill I have made. Next I just barely cover the seeds with fine soil and with the back of my hand or fist firmly compact the earth on top of the seeds I have just planted. Picking up the bowl, I take a long stride or two, squat, and seed another four- or five-foot section of the row.

For all small seeds the procedure is the same. If the soil is moist and finely prepared, there is no danger of planting the seeds too shallowly as long as they are covered and the earth is firmed down on top of them. Deep planting means only a longer lapse of time before the seedlings appear, and the sooner they come up the better, for they will have a much better chance of achieving a good stand. I recently ran across one authority who says, "It is good practice to plant seeds from one to two inches deep." If he is speaking of small seeds, and he does not specify, I could not disagree with him more violently. The essence of good seeding is to have the seeds as lightly covered as possible, and to have them placed sparingly in the row.

Each packet of seed, if it comes from a reputable grower, will have the percent of germination of the seed and the date the tests were made printed on the packet. This percent of germination should be your guide in how thick the seeds should be scattered, and in most cases you will find the rate of germination is so high that you will not have to make any allowances for infertile seed. Care in seeding will save more than twice the time taken when it comes to thinning later on. I consider it so important to have this phase of gardening done just right that, while I do have outside help at times, I will let no one plant seeds for me. Each time I have had others help me with the seeding the results have not been satisfactory. So if it seems that I am being too detailed or repeti-

PLANTING SMALL SEEDS

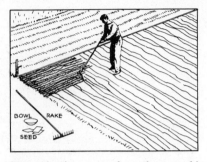 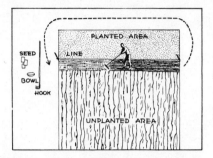

Start planting top of garden, working down grade. From line marking last planted row, start new area 3 feet wide, using potato hook, walking backward.

Pulverize new strip well to far end of garden. Carry hook around garden to starting point. Pick up rake and follow over same area, walking backward, smoothing carefully.

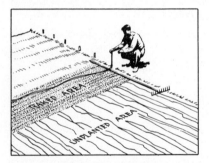 

From far end, move line to mark next row to be planted. Use rake handle for measure, notched at 12". Pick up rake, go around garden to starting point.

Again using rake handle measure, move near stake down 12" and pull line taut. Line now marks row ready for planting. Pour seed in bowl.

With bowl, take long stride out into garden, face line and squat. Reach back to stake and draw shallow furrow with forefinger of right hand resting against line.

Distribute pinch of seed down furrow. Cover and firm with fist. Repeat process to end of row, then around garden to starting point for preparation of next strip.

[86]

CLOSE PLANTING

Small seed varieties in rows 12″ apart. No need for between-the-rows cultivation once there is good growth.

tious in my seeding instructions, it is because I cannot overemphasize the importance of careful procedure in this most important step in gardening.

For the most part the procedure for planting the small seeds which I have listed is the same for all seeds. Some however, are easier to plant than others. Parsnip seeds, because they are designed by nature to be distributed by the wind, are the very devil to plant if the wind is blowing; beets and Swiss chard, because of their size, are comparatively easy; but it should be remembered that these are, in reality, small bunches of seeds, so they should not be placed too close together. Parsley is the tiniest, but a good stand will result even if the seeds are a bit

thick in the row. Salsify seeds are not actually small, and they can be placed in the row one at a time. Lettuce, endive, and carrots are difficult to get thinly spaced because of their size and shape. Radishes, turnips, and spinach are all relatively easy for their shape allows them to fall evenly when worked between the thumb and forefinger.

All of the plants under discussion are frost-resistant and so can be planted as soon in the spring as the condition of the soil will permit. In deciding which to plant first, consideration is given to the length of time to maturity, and in what part of the season they will do best. For instance, it is well to remember that lettuce seed will not germinate if the temperature of the soil is over

[87]

60°. Radishes will mature in twenty-two days and thus can be the first of the garden produce to reach the table; for this reason they should be among the first seeds planted. Spinach will be ready to pick in thirty-six days and, as it is an early season crop, it also should be among the first planted.

The first planting of lettuce should be of the looseleaf variety which matures more rapidly than does head lettuce, and thus comes sooner to the table. This can be in use while waiting for the head lettuce to mature. Parsley and parsnip seeds are slow to germinate, so it may be advisable to mix in some radish seed when planting these vegetables so that the rows will be marked. The radishes can be pulled and used as the other seedlings appear. All the root vegetables are late season or fall crops, so there need be no rush in getting these planted, but it is a good idea to have a couple of rows each of beets and carrots for early use.

Succession plantings will depend on the climate. Here in the mountains of Vermont we have no need to bother about such refinements. The season is so short that if the seeding takes place over a period of four weeks' time, the crops will be spread out to the extent the climate permits. In any climate it will be well to have the plantings spread out in time and to have the varieties separated by plantings of other varieties. Whether or not any advantage in growth will accrue from this procedure, as is claimed by some of the organic gardeners, I am not sure, but it makes for a pleasant variety in planting, and, to my notion, it improves the looks of the garden.

You will notice that I have not mentioned onions among the small-seeded vegetables, and small seeded they certainly are. Few home gardeners whom I know plant onion seeds, most of them preferring "sets," which are small dried onion bulbs that may be purchased at seed stores by the quart or pound. These bulbs are easier to plant than seed, get off to a quicker start, and do not require thinning as would be the case if seed were used. For these reasons, many seem to prefer their use to the sowing of seed, but here again I cannot agree with their thinking, I discovered that I could grow larger onions from seed than I could from sets, and I knew what variety of onion I was getting, which is not the case when sets are used. It is true that sets will produce bunching onions or scallions before seed will, but for the home gardener these can be obtained more satisfactorily by using Egyptian onions of which I will speak in more detail later. For another thing, onions grown for use as sets are grown small, that is, grown in poor soil with deliberate care being taken that they are small. This is not a good way to start off if one is growing table or storage onions. For these reasons I have gone along with onion seed and have had consistently good results in this climate with Early Yellow Globe. For the past four years I have not planted seed, however, for I find I can buy onion plants shipped from Texas air freight for a little less than five dollars per thousand, and I have my choice of four varieties. This I have discovered to be the most satisfactory method of all for raising onions, and is the one I recommend for the New England climate. If ordered in time, the plants will come early in the spring; they can be set out as soon as they arrive, so they will be among the first things to be planted in the garden. The method to be followed is the same as that for small seeds; but instead of making a shallow trench with the finger, I make holes with my forefinger, spacing them about three inches apart. As I proceed along the row, I place an onion plant in each hole, firming the earth around it.

In my garden, the first seeds to go in the ground are peas, for they will germinate in cool, moist soil, and the plants grow and produce best in the

PROGRESSIVE PLANTING

Peas, (background) well up. Radishes, lettuce and beets starting in foreground plot at left, with unplanted portion at right.

cool, damp weather of spring and early summer. Detailed procedures for setting the wire, trenching, and planting will be found in Chapter Nine under "Peas."

In order to have the crop spread out over a period of time, all the peas should not be planted at the same time unless two varieties are planted, one of which takes longer to mature than the other. This point will be discussed later. Between the first row of wire and the edge of the garden there is room for a row of vegetables which will be out of the ground by the time the peas are ready to pick. I suggest that this outside row be planted one-half to looseleaf lettuce and one-half to radishes. The space between the rows of wire is suitable for the first plantings of spinach.

From now on, any of the small-seeded vegetables, all of which are frost-resistant, may be planted, as may also those transplants which are frost-resistant. Not much will be gained if the sets which have led protected lives in cold frames or greenhouses are set out in cold, blustery weather. Cabbage, Brussels sprouts, and broccoli can stand some frost, even when relatively tender, but cauliflower can stand less so it does not pay to rush the season too much here even though early fruits are appreciated. Of these four plants, broccoli and cauliflower will develop sufficient variation in maturing so that one planting is satisfactory for the home garden. For each of these

vegetables, nature will spread out the time of harvesting, and, in the case of broccoli, after the principal heads have been cut, secondary heads will develop so that the harvest is continued on into the fall. Brussels sprouts are a long-maturing crop and will be, in this climate at least, one of the last vegetables to be gathered. It will be well to have two plantings of cabbages, the earlier summer variety in the first planting and the fall cabbages later. If the latter are planted too early, the heads will burst and so be unsuitable for storage. Celery is frost-resistant and very hardy; it can, therefore, be set out as soon as the soil is ready, although, for the sake of convenience, I always set it out at the same time as the others.

Since ordinarily I plant all the seeds for my plants in the cold frame at the same time, they will all be ready for setting out at about the same time, but it will not matter at all if these settings are spread out over several weeks. My practice is to sow the seeds in rows four inches apart as rapidly as is compatible with good speed. These rows will have to be kept free from weeds and the space between the rows can be cultivated by a small hand claw, an instrument which should be included among the garden tools, along with the trowel and the dibble.

When the time arrives, trowel up some forty or fifty plants at a time discarding those which are too tiny or those with poor root growth or

HOW TO PLANT SETS

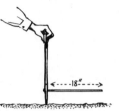

Wait for a gloomy day. Procure the flats, prepare the soil and set the line.

Mark off 18″ intervals along line, with rule or hand-made transplant divider described in text.

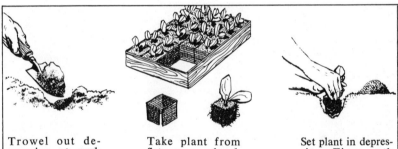

Trowel out depression at each interval.

Take plant from flat—remove band.

Set plant in depression. Firm earth around it.

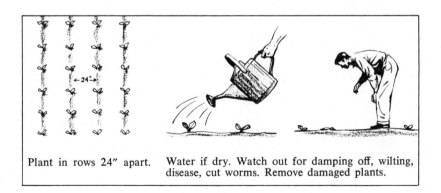

Plant in rows 24″ apart.

Water if dry. Watch out for damping off, wilting, disease, cut worms. Remove damaged plants.

dampened-off stems, and get them set out with the minimal time lag between the digging out and the setting in.

When it comes time to get the sets in the ground, I watch the weather. With the first overcast day, preferably with mist or light rain falling, I start setting them out. If the weather can be caught at the right moment, the whole procedure is simplified for it will then not be necessary to water the plants as they are placed in the ground. If the weather continues lowery and wet for a few days afterward, there will be no need of watering the plants after they have been set out; otherwise, they will have to be watered until the roots get a good start. Wilting of the plants gives a clue as to the need for water.

Be sure the soil in the cold frame is well soaked before removing the plants, and that the flat which is used to transport them from the cold frame to the garden is moist. Do not let the tiny roots dry out. Never dig up more than can be planted before a drying of the roots occurs.

The weather being right, start the operation by first setting the garden line, spaced thirty inches from the nearest vegetables, or, if at the outside of the garden, twelve inches in from the edge. For the sake of convenience, I space cabbage, cauliflower, broccoli, and Brussels sprouts all the same in spite of the fact that cauliflower and broccoli take more space than the other two. If the soil is rich the plants do just as well, and the only questionable result is that care will have to be taken in getting through the rows when the plants have arrived at their full growth.

A tool such as is pictured can be devised, but with the passing of years I have discovered that using a dibble is more convenient and speedier as well. A dibble can be bought at the garden supply houses and it consists of a conical shaped steel tool attached to a suitable handle. The pointed end is thrust into the soil thus making a hole for the plant, and the tool as a whole can be used for measuring off the distance between the holes. The dibble I use is just one half of the required distance between plants. That row finished, make a mark twenty-four inches from the line stake and set the stake firmly in the ground at this point. Move the line stake at the other end of the row, taking care that the line is taut and exactly parallel to the last row, and start all over again. For celery, the procedure is slightly different. The weather conditions should be the same and the same row spacing will do, but the distance between the plants is much less. Celery plants can be set five or six inches apart in the row, and a mark on the dibble used will make even spacing an easy matter.

Tomatoes and peppers are not frost-resistant and cannot be set out until all probable danger of frost is past. Tomato and pepper plants are in a different category entirely from the others. These are extremely sensitive to frost, can bear none, in fact, and in northern climes the plant will have to be started in the house or else purchased from a grower. I prefer to grow my own and so I prepare two standard sized flats, covering the bottom on the inside with several layers of newspaper, then filling with fine compost. These I place on a suitable table which in turn is covered with newspaper and placed in front of a south-facing window. I plant the carefully spaced seeds therein during the first week in March. There are other methods than this for raising suitable plants as any seed-house catalog will describe, together with the appurtenances thereto.

As soon as the weather will permit and preferably before the plants are more than six inches high, they will have to be transplanted to the cold frame, and this, of course, only after all danger that a cold night will penetrate into the confines of the frame, for after one such, the plants are done for. I go to some pains to avoid such a loss, rigging

Broccoli, brussels sprouts, cabbage and cauliflower planted from sets about May 31.

[93]

up a couple of 100 watt electric light bulbs inside of the frame, and making provisions for covering the glass (or plastic) with a blanket.

After all the danger of frost is past the plants can be set out in the garden. But before this, some pains must be taken to see that they are hardened. In other words don't plump them into the garden without an interval of having them gently exposed to cool weather. If you buy them procure good, healthy tomato plants, and, if you want early tomatoes, the plants should be in

blossom or even have small fruits started when you set them out. Place the trellises which have been prepared in advance, preferably in a warm place well protected from the wind.

In order to get the most out of both the soil and the garden it is important to devise a system of inter-planting. For example, I make it a practice each year to plant spinach in between the rows of peas. We grow quantities of spinach, for organically grown spinach free from side dressings of soluble ni-

Same view as preceding page, 45 days later.

[94]

trates, which indubitably impart a bitter taste to the leaves, is delicious as a salad. Since we supply large quantities of salads to two nearby gourmet restaurants, great amounts of spinach are used thus. Perhaps a small supply of spinach will be deemed sufficient in the kitchen garden. But if you down quantities of spinach use the space between the pea rows for it. Otherwise radishes and loose leaf lettuce can be planted here, for these crops will all be out of the ground before the peas are ready to pick.

Another place for inter-planting is between the rows of cabbage, etc., and around the hills of squash. Here will be found plenty of room for vegetables which will soon be out of the soil. But these arrangements are dependent upon the uses to which the garden is put. In any event it is foolish to let good space go to waste. Use it to its best advantage.

Another trick which I have developed in recent years is to plant my cucumbers along the outside rows of corn. The vines can be directed inward and will often climb the stalks. The combination of corn and cucumbers seems to be a happy one.

But let me emphasize that the mark of the expert is high production per unit of soil. In 1970 our garden produced vegetables having a cash value of $1,493.83, and the space involved was 8,352 square feet, which is less than one-fifth of an acre.

Now we are left with only those vegetables which are frost-tender; as a consequence, these will be the last seeds to go in the garden. This group includes corn, beans, squash, and cucumbers. Inasmuch as the planting techniques for each are different in some respects, generalized instructions are not particularly useful. Therefore we will omit any discussion of the planting of the seeds here; specific instructions for planting will, however, be found under the proper heading in Chapter Nine.

CHAPTER EIGHT

Cultivating the Garden

WITH GOOD SOIL, good seed, and proper planting, the garden will be up and off to a good start, and then will come the necessity of caring for it. The number of poorly cared for gardens seen about might lead one to the conclusion that this is a formidable task, but such certainly is not the case. If the size of the garden is within our means as far as expendable time is concerned, and we are on hand to give it care when care is needed, the results will measure up to all expectations, and the demands on our time will not be excessive.

The first step in saving time and energy later is to cultivate each row as soon as the line of seedlings is visible. If the ground has been worked in preparation for the seeding as has been described, the plants will be off a whole jump ahead of the weeds. All we have to do is to maintain this head start, and to accomplish this the first cultivation must be given as soon as it is possible to do so. To delay now means to lose our advantage. The expres-

sion "to grow like a weed" has evolved out of sound human experience, and if we but give the weeds a chance they will assume the lead in the race for growth with such vigor that to head them off becomes an almost impossible task. This is the stage of gardening that produces the biggest crop of discouraged gardeners. Once the weeds have overcome the handicaps imposed on them in the beginning, the task of controlling them becomes increasingly difficult, so that the time we spend cleaning out one row will be used by the weeds to do permanent damage to the plants in the next row. For these reasons the first cultivation must be given at the earliest possible moment, and to all rows at the same stage of growth at the same time. Now we can see the importance of not planting all the seeds at one time, but of planting a few rows of one variety, then a few rows of another, and of mixing radish seed in with the plantings of slow germinating varieties. This first cultivation can be given very quickly by

Cultivating late summer lettuce.

using the potato hook as described in a previous chapter, moving backwards astraddle of the row, working with care close to both sides of the row in the middle, and close up on one side of each of the two outside rows. In this way two rows are done at once, and after a bit of experience it can be done very quickly, effec-tively putting the weeds out of competition. At this stage of the game, weeds are tender and more easily killed that they will be later; therefore, this, the most important of all cultivations, must take place just as soon as it is possible to trace the faintest line of seedlings.

Cultivating the garden serves two

[97]

purposes: the first and perhaps the most important is to control the growth of weeds. In addition to this, the condition of the soil is affected making for better plant growth. Loosening the surface during moist weather enables the soil to soak up greater quantities of water, and in dry weather the loosening of the soil at the surface forms a dust mulch, which prevents excessively rapid evaporation of water from beneath. In any event, frequent stirring of the soil leaves it in what Rawson describes as a "fresh and lively condition."

LATER CULTIVATION

Straddle the row and move backward, working two rows at once.

[98]

After the first cultivation has been effectively accomplished, conditions of weather and weed growth will determine when the next one must take place. If dry weather succeeds the first cultivation, there may be no need for a second going over until it rains, unless the surface has become caked in the meanwhile. Whatever the weather, the rows should be gone over again in the same manner as before as soon as the weeds show any signs of reappearing. This time the cultivation can be deeper than before, for the seedlings will have become more deeply rooted and the danger of ripping them out or disturbing them unduly is past. The best time for the second, or any subsequent cultivation, is after a rain. Cultivation before a rain is apt to be pretty much wasted effort, for the rain falling on the uprooted weeds will give them new life, and they will take root again unless removed from the rows. This is particularly true if the potato hook is used for cultivation as I recommend, for in this instance the weed stems are not cut off; they are uprooted instead, and, if the sun does not kill them, which is normally the case, they must be raked off the plot and placed on the compost pile. Here is where the narrow rake comes to use in cultivating the small-seeded vegetables, the rows of which are only ten or twelve inches apart. Some species of weeds have more vitality than others, and the gardener will have to exercise his own judgment in deciding whether or not the weeds must be raked off. I find that normally such a procedure is not necessary following the first and second cultivations. With moderately decent luck in the weather, no more than three cultivations should be necessary, for by this time the growth of the plants should be such that the tops are beginning to close over the rows, and subsequent cultivations will do more harm than good. In the case of onions, however, the need for cultivation will continue to the end, for the stalks produce too little shade to be effective.

While it is important to cultivate after rain, this should not take place while the plants are still wet and the soil moist and soggy. Plants are tender when wet, and vulnerable to mildews and rots if bruised or broken. This is particularly true in the case of beans, and it is deliberately asking for trouble to molest bean plants in any manner at all while the leaves are wet. Wet soil will not break up as finely as it will when slightly dried out, and it becomes more tightly compacted by walking on it than does drier soil. For these reasons it is best to keep out of the garden while it is still wet.

In an earlier book I expressed convictions concerning the necessity for watering the garden which I do not hold today. Further experience, and the urgings of common sense have led me to the conclusion that it is desirable to provide some means of irrigating one's garden if it is to be the best possible. Since this locality is one where droughts do not occur, and in nature's way sufficient moisture will ordinarily be supplied to insure good growth when coupled with proper practices of cultivation, I, for years, made no provisions for watering, though I did admit that "there have been one or two occasions when I would have watered if facilities for doing so had been available." So now I must withdraw from that position as gracefully as I may.

Most garden crops, be they leafy, or roots, or fruits, consist principally of water. And if a cucumber is, we will say, ninety percent water, how can one expect to obtain perfect fruits if at any time there is a lack of moisture? In fact cucumber fruits, in their shape and size offer a convincing record of the amounts of water which have been available to it. Cucumber growers will all have noticed a fruit small at the stem end which suddenly swells out to normal size, or one which starts out nobly only to be squeezed into a miserable pointed end, or even those which swell and shrink and then swell again, thus giving visible evidence of the dry and rainy spells. Good fruits and leaves and roots will result if plants are never allowed to suffer from lack of water, and the measure of this is the condition of the soil. As soon as the soil in the garden becomes dry on the surface it is time to water, and if the water is supplied at this early stage there will be no

necessity for heavy doses. Nor will that old bugaboo of superficial root growth ever become a reality, (if in fact such ever is the case). This I am inclined to doubt, but in any event, if the surface is not allowed to become parched, the amount of water necessary will be held to a minimum.

In order to be able to provide this care, some sort of a system is required, for, except if the garden is a real tiny one, the gardener cannot afford the time to stand holding the nozzle of a hose while the ground soaks up the water. No home vegetable garden needs to have any more elaborate means for irrigation than an outlet or sill cock somewhere near the plot, and enough hose to reach every part. This hose is attached to a rotary type lawn sprinkler whose base is no more than six inches across. This base is then fastened to a narrow strip of sheet metal, wide enough to receive the base, which is bent up at the front end. Preferably this turned up front should be high enough so that by means of a hole cut in it the hose can be passed through and then attached to the union on the sprinkler.

Now place the sprinkler between the far end of two rows far enough in from each edge of the garden so that all parts will be well within its range, and turn it on. When this section has been soaked, standing at the near end of the rows, pick up the hose and pull the sprinkler along, positioning it now at the edge of the area already sprinkled. It will be necessary to have a movable stake (I use a ⅜ inch iron rod) set in the edge of the garden at the near end of the row, so that the hose will not drag across the rows as one proceeds down along the rows to put the sprinkler in its original position between the rows.

One of the important phases of caring for the garden is thinning. All of the small-seeded vegetables with few exceptions must be thinned in order to have a good crop of prop-erly matured vegetables. Use reasonable care in seeding parsley and there will be no need of thinning, for there will be plenty of sprigs of usable parsley even if the plants are close together. The large size of salsify seed makes it possible to plant these seeds so that there need be no occasion to thin the plants. I seldom deliberately thin looseleaf lettuce for we start using it on the table as soon as the leaves are an inch or two long. Thus, in effect, this variety of lettuce is thinned as it is used. The same

Interplanting of spinach between pea rows ready for thinning. Plants should stand 3″ apart.

system can be used in home gardens with spinach if great care is used in planting. Call it thinning or call it harvesting; the first several messes for table use will be in the nature of thinning for use and will leave the rows in proper shape for full growth of the plants which remain. To achieve maximum growth, spinach plants should be about three inches apart in the rows, and looseleaf lettuce, four or five inches. Radishes will not require any real thinning either, for they can be planted sparingly and the first pickings of small radishes can be such as to leave room for the plants which are left in the row. When it comes to beets, real thinning must take place, but the thinnings are saved and used on the table. There is no dish much more delectable than the first beet greens of the spring. Here again the wisdom of not putting all the plantings of a given vegetable in the ground at the same time becomes apparent, for a preliminary thinning of the first two rows of beets can be made before the next two rows are in need of it. The second and final thinning will result in a mess of baby beets, pea-to-walnut-sized beets which are cooked along with the tops and which make a dish fit for a king. This final thinning should leave the beet plants two to three inches apart in the rows.

Head lettuce must be thinned as soon as the plants are a couple of inches high, leaving just one lusty plant every eight or ten inches. Now there will be a surplus of thinnings, of which some can be used for salad, and some as transplants, if there is space for them in the garden.

When it comes to carrots, a real job of thinning confronts us. Carrot seed are such that it is difficult to get them thinly spread and evenly spaced, and carrots will not do well if left close together without room in which to expand, and, if touching, the roots are liable to wrap themselves around each other. Carrots should be spaced one-and-a-half to two inches apart in the rows. In a home garden the thinning can be done in two stages, the first, when the plants are three inches high, thus cleaning out the worst of it. The second thinning can be done after the roots have achieved the size of a pencil, and the thinnings can then be used on the table, for these baby carrots are too good to be missed.

If onion seeds are used, thinning will be necessary. This can be done after the plants have grown to small scallion size, and the thinnings can be used on the table. They should be culled to stand two inches apart in the rows. Parsnips might just as well be thinned while the plants are small and easy to handle, as there is no use for the thinnings that I know of. The plants should stand about three inches apart in the rows. Turnip thinnings can be used as greens, and probably rutabaga greens are good, too, but there will be such a spate of thinnings to be had at this time that undoubtedly most of these greens will have to go to the compost pile. Turnips should be thinned to two to three inches, and rutabagas to three to four inches.

Endive should be treated the same as head lettuce, and Swiss chard should be thinned to stand about four to six inches apart in the rows, the thinnings being used as greens.

In every instance just described, weeding should take place as the rows are thinned. If the between-the-row cultivation has been carefully done, there will normally be little to do in the way of weeding, but in any event, it is important that the rows be clean of weeds.

As soon as the shoots of corn are up two or three inches, they should be hoed and hilled up slightly. After this it should not be necessary to go over it more than two additional times, hilling up a bit more with each hoeing. When the corn is knee high, it will no longer need any cultivation.

Keep the soil loose and free of weeds around the small plants of

Corn patch hilled up slightly with first hoeing. In background, beans at right, cabbage patch at left showing a month's growth.

squash and cucumbers, and as they grow out and cover more ground, keep the surrounding area as clean as possible of weeds without disturbing the plants. Keep the peas free of weeds, hand weeding if necessary while the plants are small, after which little will be required in the way of cultivation. The sets, such as cabbage, cauliflower, etc., will take about three workings before the plants are big enough to shade the ground and make further cultivation unnecessary.

In my garden there is small further need for cultivation after the middle of July; the thinning has taken more time than has the actual cultivation with the potato hook; but make no mistake: if the cultivation had not been done promptly when needed, there would be little necessity to spend any time thinning, for there would be little or nothing to thin.

One phase of garden care which often assumes great importance is the need for counteraction against the attack of insects and disease. To be sure the need for defense against these enemies is always at hand, but it is much easier if the plants supply this defense themselves than it is if we have to wage an active campaign. The need for spraying and dusting has come to be such a commonly accepted part of gardening that it is hard to swallow any statement which says such operations are not at all necessary, and yet that is exactly what I must say. For over twenty years now I have neither dusted nor

sprayed any plants in my garden, and losses from the ravages of insects or disease have been negligible. If nature is in balance, or as near as it is possible to have it so, in the environment of a healthy garden, plants will resist disease, and birds and some bugs, such as ladybugs and praying mantis, will feed on other bugs to the point where only a nominal amount of damage is done to the growing plants. This has been my experience and I can see no reason why this same experience cannot be shared by others.

If the soil is rich in humus and in good tilth, with nothing in the way of chemicals added to disturb the action of nature, the healthy plants will resist disease in a normal fashion and the bug population will be under control. In any event, insect damage can be held to a minimum by close observation and careful handpicking. In the spring, while working the soil and planting, cutworms will be uncovered. Get to know them and kill them on sight. If in the morning you see that a plant has been chopped off, dig shallowly around the base of the plant, uncover the worm and kill him. The various beetles which attack beans, squash, and cucumbers will be on the leaves of the plants in the morning, and before the sun strikes them they will be so sluggish that you can pick them up and crush them between your fingers. Make a tour of inspection for a few moments each day during the egg-laying period and destroy the egg clusters as you find them on the undersides of the leaves. The few larvae that do appear are easily killed between the fingers as they reveal their presence by lacy spots on the leaves. Encourage the birds to stay around the garden site; in fact, encouragement will not be necessary, for the presence of insects in itself will attract the swallows, kingbirds, and other bug-eating birds. All the gardener needs to do is not to discourage them, and the presence of house cats will do just that. Cats and gardens do not go well together for the presence of cats will tend to keep the birds away and birds are the principal control against an overpopulation of insects. Nature's balance is a delicate thing indeed, and a disturbance in one direction may show up as an imbalance in a totally unexpected direction. I must tell again the true story of the skunks, the turtles, and the ducks.

On a farm in New York State there was a small pond which, in the fall of the year, afforded some duck shooting to the owner of the place, who loved it, and who looked forward each year to the prospect of having some sport in the season when the ducks came in. Year after year he was not disappointed; but then one year no ducks came in to the pond, nor did they in succeeding years. The puzzled and disappointed duck hunter called in the services of the conservation department to help him solve the mystery. After careful observation and study, it was dis-covered that the following sequence of events had taken place as the result of the farmer's two sons having grown to the age where they like to hunt and trap. There were skunks on the place and, as the pelts were worth a few cents each, the boys took to trapping them. This commercial activity on the part of the boys resulted in the cleaning out of the skunks. With no more skunks to hunt out and eat the turtle eggs, the snapping turtles which lived in the pond throve and multiplied pro-digiously and so infested the pond that the ducks would no longer come there to breed or feed. So in this case you kill a skunk and lose a duck; in the case of the garden, you keep a cat and lose a cabbage.

As far as disease is concerned, see that the plants are in healthy and vigorous growth; avoid bruising or breaking them, and keep out of the garden when it is wet. As for the bugs, be vigilant and unrelenting in your search for them and let nature take its course, undisturbed by you as far as possible. A primary re-quisite for healthy plants is healthy seeds, and so it is of the greatest im-portance to get the best seeds possible. It will not pay to save a few cents by buying seed that you cannot be sure of. Get the best seed from tried and true seedsmen. To order direct from the producer is the best plan, and be sure that the seed con-tainers are marked with the results of germination tests and the date the tests were made.

[105]

As I mentioned earlier, I do use wood ashes in my battle against insects and disease. A long while ago, when troubled with maggots in my onions and cauliflower, an oldtimer told me to use wood ashes, a good handful around the base of each plant, his theory being that the flies would not lay their eggs in the wood ashes and so the maggots would not develop. Whatever the reason, the treatment worked, and I discovered that in addition to this plants treated with wood ashes seemed to gain immunity to club root. Following up this line of inquiry, I discovered that wood ashes would also control scab in beets and turnips. So it is that I use wood ashes on onions, cabbage and all that family, and on beets and turnips as well. As I said before, I do not know whether the beneficial results come from direct action on the disease and insects, or whether they come from adding elements to the feeding of the plants which increase their resistance.

If it seems too hazardous to start off making no provisions for the battle against bugs and disease, dusting with rotenone and pyrethrum will help, with no resulting harm; at least they are not poisonous to humans; so, while the presence of insecticides is not recommended as a dressing for green vegetables, these will be found to be less poisonous than Paris green, for instance. A more detailed description of the enemies of each variety will be given in the next chapter, wherein each vegetable will be discussed separately.

This chapter on the care of the garden should not close without a discussion of some of the common weed nuisances, or at least the ones with which I am familiar, being present in my garden.

One of the worst offenders is sweet clover, and there are many places where the presence of this plant would be welcomed. Even in the vegetable garden it has one redeeming feature, for clover and in fact all legumes, have the peculiar property of transforming atmospheric nitrogen into compounds which are available as plant food. Certain bacteria invade the roots and nodules are formed wherein these nitrogenous compounds are stored, so that the clover plants store up available nitrogen and thus enrich the soil in which they grow. There is an old saying around here that wood ashes will grow clover on a bare rock, and I suspect that being cursed (or blessed) with clover in the garden is more or less directly the result of the liberal use of wood ashes. In any event, it does constitute a nuisance for it grows rapidly and spreads and will crowd out the vegetables. When young, it is not much trouble to cope with, but once it gets a good start, the incredibly strong roots will reach deep in the earth making it the very devil to get out. It is so strong that if any of it is left to get a good start it constitutes a major hazard to the light tractor with which I plow in the fall. Its grip on life is amazing, and it will continue to grow if chopped up or turned under, so the only way to control it is to pull it out before it gets too much of a start, and place it on the compost pile.

Worse than clover to start out with, but becoming less of a major problem each year, is "witch grass" or "quack grass." This member of the grass family (*Agropyron repens*) propagates by creeping underground stems as well as by seed, and it is this first characteristic which gives the gardener the most trouble throughout the eastern states and as far west as Minnesota. If this pest is in residence around the garden site, and it generally is, for it thrives in rich, well-drained soil, it will send its long, horizontal, underground stems out into the garden from the edges and paths. The tips of these runners are tough and sharp; they will easily pierce a large potato; but the stems themselves are tender and break easily at the joints, making it almost impossible to pull it out without breaking, and each joint or section left in the ground will start a new plant. Repeated clipping with the lawnmower seems to discourage this pest in the lawn, but somehow or other each spring there will be a new crop of it invading the garden from the paths and edges. Once it takes over completely, the usefulness of the site as a garden plot is over, so it must be kept under control. To accomplish this, I find it necessary to hand spade a two-foot strip all around the outside edges of each of my garden plots. The procedure is to set the garden line along the edge of the garden, cut down deeply along the line with a sharp-edged spade, and then, with a fork, spade up the whole strip, carefully shaking out each forkful so that every piece of "quack grass" stem can be picked up and thrown aside. This refuse should then be raked up and destroyed. If thrown on the bottom or in the middle of the compost pile, it will decompose, but if near the sides or top it will take root and thrive, so perhaps the safest thing to do is to burn it. Inside the garden proper, whenever and wherever a shoot appears at the surface, dig it out, following along each of the stems until every bit is removed, even though some of the vegetables are destroyed in the process. Do not underestimate the power of this enemy; once the upper hand is lost, the battle is a losing one.

There are only three other weeds which appear in my garden to the extent that they pose any problem, and in contrast to the previously described two, which are perennials, these three are annuals. They are chickweed and purslane [also known as pusley] and frenchman's weed. The proper control of all annuals is to kill them before they have a chance to blossom and go to seed. Weed seeds will be present in the garden soil in spite of all precautions; they come in the manure; they are brought by the wind; and come

WITCH GRASS

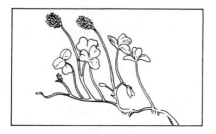

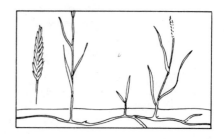

SWEET CLOVER

they will, but at least we can refuse, to grow seed plants right in our own garden. For this reason all weeds in the garden must be kept down, not merely those whose presence interferes with the growing vegetables. Never let any weeds in the garden, or in the vicinity of the garden for that matter, arrive at the seeding stage if it is possible to prevent them. This is the first step in the control of annuals; for the rest, keep the garden clean and well cultivated. The reason that chick-weed and pusley persist from year to year is that they are tough and hard to kill, and while they are uprooted easily, many of the plants will take root again unless removed from the ground. Both of these weeds grow close to the ground, and both have small flowers and seed early. These characteristics enable the weeds to persist from year to year, but actually they do not constitute a major problem of control unless they are allowed to get out of hand.

The onset of frenchman's weed in this garden is a recent one, and it turns out to be the worst one of all. My prayer is that you, gentle reader, do not know what it is, and that it is never your cruel fate to learn. Frenchman's weed comes late in the season and it will literally take over if it is not controlled. It is extremely frost tender, but it is a vigorous grower, and even though it is no problem up until about the end of July, what it can do between that time and the first frost will astound you. Earlier I have said that there should be no need to hoe corn after it becomes knee-high, but if you have frenchman's weed in your garden you can forget that. After a certain height it becomes impossible to hoe corn, but until that time comes it is necessary to keep this most evil of all weeds cut down. After that there is nothing one can do, and by the time the ears are ripe and ready to pick the weed will be knee-high. I must say though that there is no noticeable harm done to the crop of corn. The same thing will happen in the bean rows, and here, even though the roots of the bean

CHICKWEED

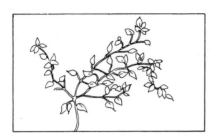

PURSLANE

plants will be disturbed, it will be well to pull out the weeds. The worst aspect of this villainous plant is that it has a vigorous root growth in a clump just beneath the soil, and so when pulled, a great gob of soil comes up along with the root. At any rate it is very tender, and serves some purpose in nailing down the soil, and when turned under in the fall, serves as green manure.

The care of the garden through the growing season is exacting; it imposes on us the necessity of doing the right things at the right time and in the best possible manner, but it need not be laborious or time-consuming. The difference between carrying this phase of gardening through with ease and pleasure, and meeting with frustration and defeat, is slight, and so it behooves us to take extreme care and to exercise our intelligence if we want to produce the best results and have the most fun doing it.

In all instances thus far I have given an explicit and detailed ac-count of the techniques that I use in preparing the seedbed, planting, cultivating, thinning, and in making compost. These procedures are sometimes at variance with those given by tried and true authorities, but in each case they are the result of my own experience spread out over a long period of time, and the use of them results in a uniformly fine garden year after year. That they are the best I know of, you may be sure; that they are the best possible is not improbable.

Different soils and climatic conditions are certain to impose modifications upon any set of agricultural practices. Each individual gardener will develop his own little tricks of the trade, and perhaps arbitrarily adopt personalized procedures which to him seem to be the best possible. I have given detailed accounts of my procedures, for they work the best of any within limitations of soil and climate. I am certain that their use should produce uniformly fine results.

A SUMMARY OF GARDEN CARE

Begin when seedlings first visible. Keep at it until rows close over.

Thin and hand weed at same time . . . water by soaking.

Handpick insects, egg clusters . . . kill cutworms . . . wood ashes for maggots.

Wood ashes control club root and scab . . . keep out of garden when wet.

CHAPTER NINE

Garden Vegetables

THIS CHAPTER WILL BE a series of sections, some very brief, each of which will have to do with a separate vegetable. There are two dozen of these items which we wish to include in our New England vegetable garden; some are of minor importance, but most of them we will definitely choose to grow. In each instance there will be data and information concerning the vegetable under discussion, including pro-

With close planting, vegetation almost closes over from row to row.

cedures for seeding, cultivating, and harvesting. Some of the procedures will have already been described in other parts of the book. For this repetition, I beg the reader's indulgence; for the sake of compact reference, however, this repetition seems necessary.

These sections will be taken up in alphabetical order, and at the end there will be a brief discussion of a few perennials which are nice to have but which require separate garden space of their own outside of the main garden area.

The inclusion of flowers within the vegetable garden space oftentimes is a sensible thing to do, particularly those for cutting or for sale, as long as their presence does not interfere with the main objective. I normally include sweet peas, but we will not permit these minor deviations to divert us from the principal task, which is that of growing vegetables. And so to beans.

BEANS

Two familiar relatives in the legume family hold the greatest importance to the vegetable gardener —peas and beans. Sweet peas and edible podded or sugar peas are pretty much frost-resistant, and so we start planting them just as soon as the earth can be worked in the spring—even up here in Vermont.

Growing beans is an entirely different story. Peas like cool weather and moist soils, and because of this they are not successfully grown everywhere, especially in places where beans do best. Beans like warm dry soil, and will be killed by the slightest touch of frost. All varieties of them, that is, except the fava bean, which is frost-resistant, but not recommended for the family kitchen garden.

The tender-podded kidney beans, native to South America and commonly referred to as "snap" or "string" beans, are the most popular of the home-garden varieties, with the possible exception of the lima bean (which can not be successfully grown in this climate). Kidney beans come in two sizes, tall and dwarf, as well as in assorted shapes and colors. An extremely interesting book written by Madame Vilmorin-Andrieux of Paris, called in its English edition (1885) *The Vegetable Garden*, a tome of over six hundred pages, lists 105 varieties of kidney beans but mentions neither of the two varieties of pole string beans with which I am most familiar, the "Kentucky Wonder" and the "Romano."

The "Kentucky Wonder" has a place in my boyhood—which was a long, *long* while ago. My father used to grow these pole beans, along with limas, in our backyard garden in the city; and alas, I knew them all too well, for in those days I did not take kindly to any denizen of the garden because that was a place of forced labor as far as I was concerned. (The only part of the operation which held any interest for me was a by-product of the cedar poles which my father got from a supplier for the beans to grow on. These poles were covered with a stringy, soft under-bark which was pleasantly aromatic, and which formed the basis for quite acceptable cigarettes which we rolled in toilet paper.)

But even more binding than the entwining of these beanstalks in the sentimental memories of my early boyhood is the fact that for the past 50 happy years I have been married to a "Kentucky Wonder,"—and so it was with reluctance that finally, three years ago, I gave up on growing "Kentucky Wonders," and switched to the "Romano" bean. The reasons for the switch were three-fold: first the climate here; the cool

PLANTING POLE BEANS

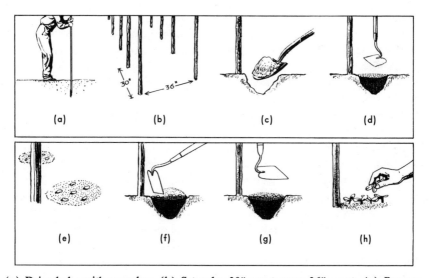

(a) Drive holes with crow bar. (b) Set poles 30″ apart, rows 36″ apart. (c) Remove one shovel earth close to pole. (d) Fill hole with compost, add soil and tamp. (e) Six beans to a hill, 1″ apart. (f) Cover with 1½″ soil. (g) Tamp down to ½″ of cover. (h) Thin to 4 or 5 healthy plants.

[113]

PLANTING BUSH BEANS

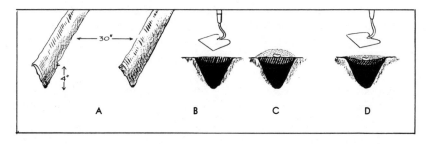

(A) Dig trenches 4″ deep, plant beans 1½″ to 2″ apart. Trenches 30″ apart. (B) Fill each trench with compost. Add soil and tamp. (C) Place seed. Cover with 1½″ soil. (D) Tamp down to ½″ of cover.

nights, seem to be more agreeable to the "Romano" than to the "Kentucky Wonder." Second they produce a crop earlier, and over a longer period of time. And third, in my judgment, they taste better. They are completely stringless and remain tender even when grown to a large size.

Right here let me comment on a fact which has to do with the dates to maturity which the growers attach to the seed varieties they sell. As in the case of the sugar pea, some quality of soil or season in my garden seems to speed up the time to maturity. According to the claims of the grower "Kentucky Wonders" are 64-day beans, exactly the same as "Romanos." But with three seasons of experience behind us, we have found that we were able to pick "Romanos" a good week before the other bean would be ready. As for sugar peas, Burpee's "Sweetpod" is listed at 68 days and "Little Mar-

vels" as 63, but in the past we have invariably picked the edible-podded peas before the shell peas.

"Romano" beans are vigorous growers—at least under the treatment we give them here—and if I could supply them with 20-foot poles and had some means of harvesting them that high up in the air, I have no doubt that the vines would welcome them! On our seven-foot poles they start back down again, and there is no telling how far they would go if left to their own devices. One wonders if perhaps this might not be the bean spoken of in the fairy tale of "Jack and the Bean Stalk."

One other comment in passing: The twisting of these bean vines (and all other vines in this hemisphere) is from right to left, or counter-clockwise. I have checked this peculiarity time and time again, and while never finding an exception, I wonder about the cause for it. But

[114]

in addition, I also concluded from the fact, that modern-day philosophers who insist that everything is relative, are quite obviously mistaken. There is a pattern in all of life, and as far as I can tell this pattern is unalterable. As for the explanation, the one which I got from an eminent astronomer was to be found in the rotation of the earth on its axis. In the southern hemisphere the direction of the twist would be clockwise. Note here that these beans, native to South America, were faced with a drastic change in their mode of life when they were introduced to northern climates.

So much for all of that; but if you are going to grow these beans on poles, it is well to know that they are very fussy about the kind they choose to ascend. The first year I tried them, about half of the hills were willing to climb; the other half simply refused to go up the poles in spite of all we would do. It soon became apparent that the non-climbers were furnished with poles which were bare and smooth, while those which climbed were planted around poles covered with rough bark. The seed growers (in this case, Joseph Harris Co., Inc., of Rochester, N.Y.) warn that the vines are not strong climbers. But I am convinced that the non-climbing is not the result of lack of strength, but simply that these vines must have a rough surface to climb on.

Actually they can be successfully grown without poles, but the yield, in our experience, was not up to that of pole-grown vines. One method we tried was to set up a row of poles about eight or ten feet apart, and stretch between these horizontal strings of binder cord about 18 inches apart, which in turn were woven together with binder cord stretched vertically from one horizontal cord to the next. Thus the beans grow on a tall fence of binder twine in much the same way that telephone peas are grown, and like peas, the bean seeds are planted in a row just under the lowest horizontal string. Planted this way, vines need some hand-training to get from one string to the next, but it can be done. Nevertheless, this method is not as successful as growing them on poles.

The bean poles should be two to three inches through at the butt, sharpened, and set at least 15 inches deep in holes made with a crowbar. Allowing a foot and a half under ground, there should be at least seven feet above ground—or more if you can harvest higher. Fortunately I can go down my brook not far from the house, and in an evergreen thicket there, cut balsam firs which perfectly fill the bill. But I should think that where such poles are not available, cedar poles with the bark on them would be acceptable to the rather fussy "Romano" beans.

Once the poles are set at least 30 inches apart in each direction, take a good shovelful of earth to the windward side of the pole (that is, on the side which faces into the prevailing wind), and replace it with a shovelful of compost. If the beans are grown thus, the wind will not be so successful in tearing the vines off the poles: and even though the garden soil may be good and rich, the compost will supply that added boost.

In the three seasons that we have grown them, there has never been any sign of disease or insect damage, but it would be sensible not to plant any beans in the same spot in the garden on two successive years. If bean beetles should appear (they have not here), my suggestion is that they be hand-picked early in the morning when they are still sluggish. A careful watch should be maintained for egg clusters which will be found on the undersides of the leaves; the beetles are more easily destroyed at this time than at any other.

I can think of nothing else to tell you about or warn you against. This is the way we have grown the "Romano" beans and they have been the most productive and succulent of all the snap or stringbeans we have ever grown. I still grow green snap beans on low bushes, and eastern butterwax beans as well, although I am not so certain now that it would not be wiser to devote more space to the "Romanos" and eliminate the bush beans which are much more susceptible to leaf wilt and mosaic disease. I suppose the answer is to be found in the poles which may be hard to get and are more trouble to install— but the results are worth all the trouble it takes.

For bush beans, no elaborate preparation before planting is necessary. Set the garden line and along it draw the sharp-cornered hoe so that there is a straight trench about four inches deep along the length of the line. Fill the trench with com-

AVOID BEAN TROUBLES

  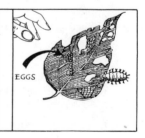

Don't cultivate when wet.

Mexican bean beetle. Handpick.

Crush egg clusters and larvae on under side of leaves.

post, cover with earth, and then drop and cover the seed before moving the line for the next row.

All bean plants are susceptible to frost, so no planting should take place until all danger is past. If conditions are favorable, the bean plants will begin to break the ground in about eight days, so if you want to cut it fine, you can plant a week before you expect the last frost. In my experience, however, there is not much to be gained by planting too soon. Bean seed will rot in cold, soggy soil and will germinate best in warm, moist soil. I do not recommend that they should be soaked before planting in order to hasten germination. If the seeds have been treated, which is usually the case, you will have lost this advantage, for whatever it is worth; moreover, the possibility of rot is increased. Nature needs no help if the seeds are planted in light, warm, moist soil and covered to the depths of not more than one-half inch of well-tamped earth. An inch or an inch and a half of loose earth will tamp down to one-half inch, depending on the structure of the particular soil. Observation will give the clue as to how much loose earth to pull over the seed. In the case of pole beans, I drop six seeds to the hill, placed in an even pattern about an inch apart. Three to four healthy plants to the hill are sufficient; plants in excess of this should be pulled out.

The row beans should be dropped one and one-half to two inches apart in the rows, and the rows should be spaced thirty inches apart.

Perusal of the seed catalog will help in arriving at a selection of varieties of beans to be planted. For bush beans I have finally come to restrict myself to two varieties only: one of the stringless green beans, either "Stringless Valentine" or "Tendergreen," the other a wax bean, either "Pencil Pod Wax" or "Brittle Wax." A home garden devoted to the production of green vegetables is not the proper place in which to raise shell beans.

To return to planting, after the soil has been prepared to receive the seed, I go along the row on hands and knees, placing the seeds carefully no more than two inches apart nor less than an inch and a half, and in as straight a line as possible. Then, with the broad-bladed hoe, I cover the seeds with about an inch or more of loose earth, which I tamp down with the back of the hoe so that the seeds lie under a half-inch of well-compacted soil. Barring a spell of cold, wet weather after the seeds are in the ground, they should break ground in seven or eight days. To plant more deeply is to delay the emergence of the plants which, in turn, increases the possibility of rot and an unsatisfactory stand of beans. After the beans are up, cultivate close to the rows without disturbing the plants, and keep the rows and spaces between the rows free from weeds until growth is such that the vegetation practically closes over from row to row. At no time should the beans be worked or the foliage touched in any way while the plants are wet, for the spores of fungus diseases are spread this way, and they can easily ruin the crop.

Be sure to take advantage of your prerogatives as a home gardener and harvest the pods just as soon as they are four or five inches long, for at this stage they are most delicious, and early picking will prolong the crop production over a considerable period of time.

If your planting produces a crop in excess of your needs, leave the vines in the ground and the beans on the vines until they mature and dry out. These beans, when shelled, make just as good shell beans as any of the varieties raised expressly for that purpose.

Beside various fungus diseases, the Mexican bean beetle is the principal enemy we have to contend with. If the soil is rich, friable, and well supplied with humus, and the plants are not worked or the fruits picked while the foliage is wet, there will be a minimum of trouble from fungi. A certain amount of damage is to be expected, but under normal conditions the loss will be negligible. As to the bean beetles, handpicking is the only control to which I resort. If you will not trust to this, you may dust with pyrethrum. The bean beetle is a flying insect about three-eighths of an inch long; it has tan wings with eight round black spots on each wing cover. These bugs hang around a week or two eating on the bean leaves. Their presence can be detected by holes in the leaves, and they can be picked off and crushed between the fingers if gotten at early in the morning, for they are sluggish at this time. Those which are not caught and killed will lay yellow egg clusters on the undersides of the leaves. A watchful eye will spot these egg clusters, which must be destroyed. Any eggs which are allowed to hatch will produce repulsive looking little monsters, yellow larvae covered with spines. It is these creatures which do the real damage, eating all except the veins and the upper surface of the leaf. Their presence is at once obvious, and as they are soft and sluggish they can easily be spotted on the undersides of the leaves and crushed between the thumb and forefinger. A reasonable amount of watchfulness is all I have found necessary to keep the bean beetle under control and, while in certain seasons there is some damage to the plants it never has amounted to much.

After the bean crop has been harvested, the vines should be pulled and placed on the compost pile. And that is all I can think of to tell you about beans.

BEETS

I know nothing special about beets except that they seem to be better liked by Europeans than by Americans. At least a Russian family buys more beets per capita than does any other of our customers. I suppose the beets go into borsch, a soup which I have never tried; be that as it may, I feel sure that the popularity of beets would increase among the natives if the fresh garden variety were the rule rather than the exception.

Beets have been picked out by the experts as being indicators of soil condition, but unfortunately the experts do not agree. One states that beets will not tolerate a very acid soil, hence a poor crop is an indication of excessive acidity, a condition which should be corrected by the addition of lime. Another says that soil preparation should be the same as for other vegetables except that lime should not be added, for this tends to form "scab disease" on beets. As I have previously stated, my experience reveals no allergy on the part of beets either one way or another. If the soil is in good tilth, well-drained, and rich in humus, good beets will be grown along with any other garden vegetable. I have used direct applications of wood ashes to young beet plants and have harvested a plentiful crop of tender, scab-free roots. I was moved to make the experiment for I had discovered that wood ashes were beneficial to turnips, and that there were fewer surface blemishes on the turnips when I did so. Wood ashes are high in lime content and are a more active neutralizer of an acid condition than is agricultural lime. So, in spite of what others may report, I insist that neither beets nor spinach nor any other of our New England garden vegetables that I am familiar with requires any special soil treatment. Once the soil in the garden has been brought to a high state of organic perfection, and the seedbed well and carefully prepared, any and all of the vegetables will do well and there need be no mumbo jumbo, scientific or otherwise, about it.

I have noticed just one thing, and that is that goldfinches seem to prefer baby beet plants to any other item of food on their menu. If I remember correctly, when I was a child on my uncle's farm in Pennsylvania, we used to call goldfinches "beet birds."

But beet seeds germinate well as a rule. Due to the fact that each ostensible beet seed, instead of being just that, is in effect a small pod which contains several seeds, plenty of plants will come up, so one should not begrudge the birds a few of them. This peculiarity of the seed should be kept in mind while planting so as not to have the seeding too thick.

Beets are a frost-resistant vegetable and can be planted as soon as the soil is in proper condition; that is, reasonably dry so that a good seedbed can be prepared. General instructions as to the planting and cultivation of small-seeded plants, as discussed previously, apply to beets. After planting, the tiny red seedlings will appear within a few days, and they should be cultivated as soon as the line of plants can be discerned. Tiny flea beetles will feed on the new leaves and at one stage may give the beets the appearance

of being somewhat fleabitten, but under proper conditions the plants grow rapidly and new and vigorous leaf growth soon makes up for the loss.

No matter how carefully planted (that is, thinly scattered in a shallow depression made by drawing the finger along the garden line, then barely covering with soil well tamped down), it will be necessary to thin the beets. This operation should take place as soon as the plants are three or four inches high. The first thinning need only clean out the thickest clumps, for it is best to make another and final thinning later. These thinnings should be carefully saved and used as beet greens. When some of the beets are the size of an acorn, the second and final thinning should take place, and the plants which are left to stand should be about three inches apart. This crop of thinnings is perhaps best of all, for the greens with the tiny, tender beets on the ends are truly delicious.

Because of the fact that all beets

THINNING BEETS

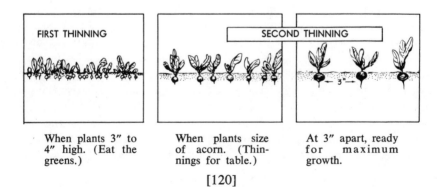

FIRST THINNING	SECOND THINNING	
When plants 3″ to 4″ high. (Eat the greens.)	When plants size of acorn. (Thinnings for table.)	At 3″ apart, ready for maximum growth.

of a given size should be cultivated and thinned at the same time, it is smart to have several plantings of beets, with not too many rows in each series, and with a week or ten days between the plantings. Thus the production of greens will be spread out over a period of time, and the thinning chore will not descend upon you all at once.

For choice of variety, again consult your seed catalog. I stick by "Crosby Early Wonder." Actually, all of the standard varieties are good, and the measure of excellence in eating is more apt to be a result of the manner of growing than of the variety planted. The growth should be rapid for prime eating and the roots should not be left long in the ground after they have attained edible size, for they will tend to become woody or fibrous in texture. For this reason I do not favor beets as a vegetable for winter use, although they can be kept over winter in adequate storage.

BROCCOLI

While broccoli is not the most important member of the vegetable family which includes Brussels sprouts, cauliflower, and cabbage, we will discuss it first because its name begins with "B." Actually the Brassica group of vegetables, of which the four mentioned above are members, also includes kale, kohlrabi, and collards, which, because I do not grow them in my garden, will not be included in the list of vegetables described in detail in this book. The Brassica group is a member of a larger family, the Crucifers, which also includes within its membership turnips and radishes. It is interesting to note that all of these vegetables are enough alike in one respect, at least, so that there is one insect pest common to all. This is a fly, the actual identity of which I have been unable to determine. It lays its eggs in the ground near the stem of any one or all of the vegetables of the Crucifer family. The eggs hatch into maggots which promptly wiggle their way into the stems of the plants in question. If the infestation is severe, the maggots will kill the plant. The northeastern part of the United States is the happy hunting ground for these nasty little monsters, so it will be well for us to take note of them. Since they cannot be ignored, I will discuss them once

Broccoli.

and for all here under broccoli; but do not be misled, maggots will attack all members of the Crucifer family, and some of them are more vulnerable than broccoli.

I have spent hours watching in the garden in the spring trying to identify the fly that lays the eggs, and, while I am suspicious of a bluish-colored fly about the size of a housefly, but much slimmer, I cannot be sure that she is the culprit; but in any event this insect, *Hylemmya brassicae*, is a cousin to *H. antiqua* which breeds the onion maggot, of which we will hear more later. To combat them, I have found that wood ashes are highly effective; in fact, I believe

that, if used with care, they can be one hundred percent effective. In cases where the vegetable in question, like broccoli, is set out in the garden as a plant, once the transplanting has taken place and the roots have had a day or two in which to become established in their new environment, take a small handful of dry wood ashes and spread it on the ground so that all the area around where the stem emerges from the ground is covered. In cases where the plant is grown from seed directly in the garden, as are radishes and turnips, spread wood ashes continuously along the row as soon as the plants have their second leaves.

It will not matter if the ashes get on the leaves of the plant; the objective is to cover the ground with ashes next to the plant stems. If high winds or heavy rains disturb the layer of ashes, I suppose it would be wise to give the plants a second application. As nearly as I can tell, the early part of the season is the worst; at least, later plantings of radishes do not seem to be as badly afflicted as do the earlier ones. So much for maggots. Now to get along with our broccoli.

Here we will start as I do, with the plant as it comes from the greenhouse or from our own cold frame. (See "Planting The Garden.")

The broccoli sets come in "flats," which are low-sided wooden boxes about sixteen by twenty-four inches. Each flat contains from sixty to one hundred separate plants, each in its own plant band. The plant bands are square pots, so to speak, made of thin sheets of wood and are about two inches square by two and a half inches high.

Transplanting should be done when the earth is moist and the weather rainy, or at least lowery, so pick a day sometime soon after Decoration Day, which fills the specifications, take your flats out to the garden edge, garb yourself in mud and water-resistant clothing, and get to work.

To simplify matters, I set out all members of the Brassica group at the same distances, the rows thirty inches apart, and the plants eighteen inches apart in the rows. The procedure for setting out the transplants has already been described in Chapter 7.

Once the plants are in the ground, they must be watered if the weather gives trouble and the plants begin to wilt badly. As soon as the broccoli plants have taken hold, give them the wood ash treatment as previously described and keep the ground around and between the plants cultivated with the potato hook so that it is always light, loose, and free from weeds. The leaves of the broccoli plants will be ravished to a certain extent by cabbage worms, but these worms will not feed on the heads; thus they will cause but little damage. The cabbage worms can be controlled to a certain extent by hunting out and killing the leaf-green-colored worms as they are discovered.

As soon as the first heads are well formed, and before they begin to loosen up and before the buds start to show the yellow of the flowers, they should be cut. Secondary heads will form along the side stems after the main head has been cut, and while these heads are smaller than the first ones, they will continue to grow long after the first cuttings have been made.

Because of soil-borne diseases, notably club root, a fungus disease which causes the root system to ball up into a club-like monstrosity, broccoli, in common with all of the Brassica group, should not be

BRASSICA CULTURE

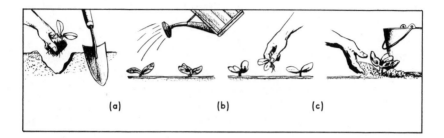

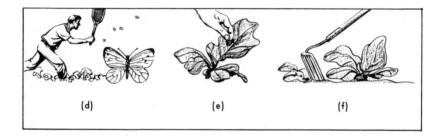

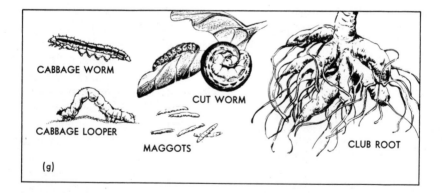

(A) Plant as in chart for "sets", Chapter 7, water when dry. (B) Remove diseased, broken plants. (C) Side dress with wood ashes. (D) Combat the cabbage worm moth. (E) Hand pick egg clusters. (F) Cultivate, keep earth loose. (G) All cabbage family plants have these troubles in common.

planted in the same soil on successive years.

After the last of the usable blossom sprouts have been cut, the plants should be pulled and placed on the compost pile. This *coup de grâce* may not be delivered until well after frost, as the blooms will continue to come along, and they are quite resistant to frost.

Brussels Sprouts

Brussels sprouts belong to the same family as broccoli but actually there are few associations in my mind between the two. When I think of broccoli I think of a juicy slice of rare prime ribs of beef *au jus*, with a baked Idaho potato, and hollandaise sauce to go on the broccoli. But Brussels sprouts, alas, conjure up memories of the awful smell they generate while cooking, and visions of dismal British boarding houses. In spite of these unglamorous associations, I must admit that Brussels sprouts have their devoted admirers and even their shameless addicts. It is for these people that I raise the darn things, and I must admit that Brussels sprouts are little trouble to grow.

The procedures of procuring the plants are the same as for the other members of the Brassica group, and the plants are subject to the same hazards. Actually they seem to be the toughest member of the family, for cabbage worms will choose any of the others before they will deign to dine on Brussels sprouts. Club root and maggots are enemies, but apparently not to the same extent as is the case with the others, and in each instance the same controls apply.

It is a long-growing crop, not sensitive to frost; in fact, sprouts are improved in flavor by light freezes in the fall. The edible part of the plant is the walnut-sized miniature cabbages which grow along the main stem of the plant at the axils of the leaf. This peculiar arrangement is a source of great amazement to all who behold for the first time in their lives the Brussels sprout plant. Here is a plant which looks something like a cabbage, but not quite, and which has a whole row of baby cabbages growing out of the stem, one at each point where a leaf branches from the main stem of the plant.

The lower buds or sprouts mature first, and they should be picked before the lower leaves begin to turn yellow. In picking, the leaf stems should be broken off below the sprout, and then the sprout itself snapped off by applying pressure with the thumb. Plants will continue to produce sprouts long after the first moderate freezes of early fall and, except for the fact that the garden must be cleared for fall plowing, could be left in the garden until well after the ground itself begins to freeze. This past fall I cleaned out my patch of Brussels sprouts and piled them on the compost heap so that I could get along with my fall preparations. This took place before the plants were dead or had finished bearing, but more important works could not be held up for a few belated Brussels sprouts, so out they went. Several weeks later I went out to trim the compost pile and, in leveling it off, uncovered many of the Brussels sprout stems. To my surprise, I discovered that the Brussels sprouts had kept on sprouting at an undiminished rate of speed, so I stopped my composting and picked at least ten quarts of prime sprouts.

All of which goes to prove that sprouts are perhaps the toughest member of the family, the easiest to raise, and, if you like 'em, well worth raising in the garden.

Cabbage

Of the Brassica group, surely cabbage is king, and as such should find a place in every vegetable garden. Besides this deference to his nobility, cabbage should be grown in every home garden for the reason that it is one of the vegetables wherein the difference in taste between store-bought commercial varieties and homegrown specimens is most noticeable. Truly there is no more delicately delicious, tender treat than the first sweet heads of early cabbage.

There are five general types of cabbage: the conical-headed or Wakefield type, early cabbage; the round head, the flat head, both of which are larger and later; the crinkley leaf or Savoy type; and purple cabbage.

After an increasing number of years of experience I have finally reduced my roster of cabbages to two, the "Early Jersey Wakefield," and "Red Acre," a purple cabbage. The Savoy types do not keep well, nor is their flavor as fine as that of the "Wakefield." These latter will provide hard pointed heads all through the growing season, and the purple cabbages will supply fine firm heads all through the winter. This spring we ate our last one in April. My choice is emphatically made on the basis of flavor and flavor alone. Neither of these types of cabbage is large, and being very tender, they are seldom found in the marketplace. But they are definitely tops for flavor. Because there is a wide spread in the time of maturity between the two, I make just one planting, setting out a sufficient number of each variety. These plants go in at the same time as do the other sets of the same group. Cabbages seem to be somewhat tougher than cauliflower, and could go in earlier, but little is to be gained by rushing the season, and it simplifies matters to set them all out at once.

Once the cabbages are in, they get their treatment of wood ashes along with the rest (see broccoli), and are

A smaller variety of cabbage seldom found in the markets, is nevertheless my choice for flavor.

[127]

kept cultivated and weedfree, all of which must be considered as standard practice.

When the plants are growing they will require so little care that there is danger of forgetting them, but you cannot afford to do this. The Brassica group must be watched carefully; the gardener must be aware of all that is going on, and the appearance of the plants will give him a clue. After the plants are well started, any sign of wilting shows that something is wrong. Other signs of distress are a curling up of the leaves and a purple tinge to the edge of the leaves. At the first danger signal the plant should be examined and the cause of the trouble ascertained. It may be that you will find a broken stem, damping off, a cutworm, maggots or club root. In none of these instances is it probable that the plant can be saved, but if they are neglected the trouble will multiply. If a stem has been cracked or broken, perhaps a heaping of soil around the plant will save it. If the plant is lying on its side, or nearly so, the culprit is probably a cutworm. Get to know this enemy at sight. He is a soft, smooth worm, metallic gray-green in color, and three-quarters to one inch in length. He is sure to be lurking just under the surface of the ground close to the plant stem he had just chewed off. Dig him out and kill him.

It is at this point that the use of the cold frame for growing plants comes into its own. From the surplus supply replacements can be supplied for such plants as have become decimated.

Damping off is a disease which causes the soft, juicy part of the stem to rot away between the main root growth and the surface of the soil. This is a condition which came along with the plants, and it shows up only in the early days of their beginning in the new environment. I count on a small loss from this source and always have a supply of plants heeled in somewhere from which I can replace the sickly plant, which should be pulled up and destroyed. As the growing progresses, and until they are nearly full-grown, there may be an occasional sickly plant. These probably are the result of infestation by maggots or infection by club root, but if the application of wood ashes has been well and carefully done, there should be few, if any, of these cases. However, if they do show up, be sure to examine the plant so that you know what is causing the trouble, and then remove and destroy it.

Once the plants are full-grown or nearly so, nothing seems to bother them, but at this stage of the game the cabbage worm will start eating on the leaves. I never have seen cabbage worms kill a plant, but in the case of cabbages and cauliflower, they will damage the heads and make them unsightly. The worms will eat their way deep into the heads of the cabbages; they will likewise drill holes on the cauliflower

[128]

Cauliflower,
before tying.

Cauliflower,
tied for bleaching.

heads leaving unsightly trails of droppings all over the otherwise spotless white surface. Outside of some eating on the leaves of broccoli and even less on Brussels sprouts, they do little harm to these vegetables, but let's face it: they are definitely a nuisance and should be gotten rid of if possible.

My system of control is to kill as many of the white cabbage worm butterflies as I can. These cannot be mistaken, for they will be conspicuous as they flit over that part of the garden where the Brassica group is planted. An old tennis racket is the only effective weapon against them that I know of, for they are as elusive as the very devil. The next stage is to search out the egg clusters and destroy them. These will be found on the undersides of the leaves and cannot be mistaken. The worm must be sought out and killed before the infestation becomes heavy, for they work their way inwards where it finally becomes impossible to pick them off and kill them. They are a velvety worm about an inch long whose lovely green color almost identically matches the color of the cabbage leaf.

Fortunately all the cabbages will not mature at once, so the harvest will be spread over a period of time, the last of the early being still in the ground when the first of the purple heads begins to harden. In harvesting, pull the plant out of the ground, cut out the head with a long-bladed, sharp knife, and place the stem root and outer leaves on the compost pile. The good solid heads are the ones ready to be pulled. If left in the ground too long, the overmature heads may burst; if so, now is the time to make sauerkraut, for once split, the heads soon rot and become useless except as material for compost.

In closing, let me repeat that the Brassica group must be planted in a new place in the garden each year, and that wood ashes, used freely, will control both maggots and club root.

CAULIFLOWER

If strict alphabetical order were adhered to, carrots would properly follow here, but it seems to make more sense to keep the cauliflower with the others in the same group. Carrots will follow next.

If cabbage is king of the Brassicas, then cauliflower must be considered as the queen. It is the most delicate of the group, its seeds are the most expensive, and when the heads are of a good size, hard and dazzling white, they bring the best price of all. Furthermore, it is considered by all authorities to be the most finicky and the most difficult to grow of the entire group. I never have had any particular trouble growing cauliflower, and I believe that if the soil is in the proper condition as I have described, if wood ashes are used freely, and if the plants are watched during the early growing period, any attentive gardener, barring spells of unfavorable weather, can grow them successfully in this New England climate.

Cauliflower are vulnerable to all of the enemies of the Brassica group, which have been previously described and the required treatment given. If it is understood that the plants are somewhat more fragile and slightly more susceptible to cold and frost, no other special treatment is necessary other than these facts suggest. After the heads begin to form and as soon as they have become large enough so that they can be easily seen, they must be protected from sunlight if we are to have nice white blossoms. To accomplish this, the outside leaves are brought together and tied with a string or a bit of raffia so that no direct sunlight can penetrate to the head within. This should be done only when the plants are dry. Another possibly simpler method is to break the inside leaves so that they fold over the heads, thus shutting out the light.

The heads should be gathered for use after they have attained full size and before they begin to loosen up. Pull the entire plant when harvesting, cut the head out, together with a row of the smaller leaves, then compost the outer leaves, the stem, and the roots.

CARROTS

Never underestimate carrots; what would a beef stew or a pot roast be without carrots? And if you are growing your own fresh and tender carrots you will discover that, if peeled and split lengthwise in quarters, they will literally be eaten by the bushel. Eaten this way, raw, they make a tasty snack for any time of the day. I have never been able to grow enough carrots in my garden so that there is a surplus left over

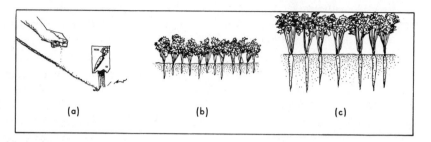

(a) You may mix radish seeds with carrot seeds to mark rows. (A *must* for parsley seeds.) (b) First thinning to ½" apart when tops 2 to 3 inches high. (c) Second, final thinning to 1½" apart when carrots pencil size. Eat thinnings.

for winter storage. All the space I can give them and the time I can devote to their care is not sufficient to supply enough of them to meet the demands of our household and the neighbors.

Carrots are a simple crop to raise, with no enemies that I know of except wildlife, such as rabbits, etc. They do require some painstaking work to bring them to maximum productivity, but the results justify the effort. First of all, the seeds are fine and difficult to spread thinly and evenly in the row, and they are slower than most in breaking through the ground, which makes it difficult to give them that first and very important cultivation before the weeds get a start. To circumvent this trouble, mix some radish seed with the carrot seed; these will come up early and will mark the row so that you can get along with the early cultivation. The radishes can be pulled as they ripen and will cause very little bother to the carrots. The

too-thick seeding is a condition that must be corrected later by thinning.

The soil and the seedbed will have been prepared as for all small-seeded crops, and the planting will take place as previously described.

With carrots, thinning is the most important operation. So that all the thinning will not have to be done at once, successive plantings should be made, two or three rows at a time, with the plantings spaced a week or ten days apart. The first thinning can take place as soon as the plants are two or three inches high. This will clear out all the excess plants and will leave a thin row wherein the plants are no closer together than a half inch or so. The thinnings themselves will be thrown on the compost pile. Later on, when the carrots are the size of a pencil or smaller, the final thinning must take place. This thinning will produce a crop of delicious baby carrots. Now the carrots in the rows should stand an inch and a half to two inches apart,

[132]

and if kept cultivated and free of weeds the tops will soon cover over the row, and all necessity for further care is over.

In harvesting do not hesitate at first to pull small carrots. Let each harvesting be a selective one, starting early with small roots; each successive time seek out and pull the larger roots. In this way the harvesting can be maintained over a considerable period of time.

There are many varieties of carrots to choose from, and again I am willing to let you make your own choice. I find no necessity for planting more than one variety, and insist that in proper soil and with proper care all varieties will be equally good. The requirements of length of maturity, storage, or the soil type may justify the selection of one strain in preference to another. I stick by the "Long Chantenay" or the "Nantes Long."

CELERY

Celery is another vegetable which may well be dispensed with in the home garden, for it is a very exacting crop to grow and, while it does not have many enemies, unless conditions of soil and moisture are just about right, the results will be discouraging. If celery is to be grown in the home garden it would be well to dig a trench along the garden line as was done in the case of peas, filling the trench with compost, but in this case the compost should be mixed with rich topsoil. The plants can easily be grown at home from seeds, but if the other sets are bought at the greenhouse, these might just as well be picked up at the same time. If grown in the house and cold frame, the plants will take eight weeks before they are ready to set out. As they are hardy, the plants can be placed six inches apart in the rows as early as the soil can be prepared, if an early crop is desired. For a late crop, set out the plants in late spring or early summer in this climate. After the stalks are well grown and nearly ready for use, they should be bleached. To do this, I go between the rows with a round-pointed shovel and throw shovelfuls up against the stalks on each side of the row. Then I set six-inch-wide

CELERY

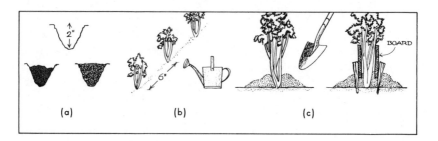

(a)   (b)   (c)

(a) Dig trench, fill with compost, mix with rich soil. (b) Space plants 6″ apart, keep moist. (c) Blanch by heaping earth against stems. Cover with 6″ boards held in place by stakes.

boards on top of the earth banking in each side, leaving most of the leafy growth protruding from between the boards, which are held in place with small stakes. The purpose of this operation is to exclude light from the stems so that they will blanch, becoming more tender and better in flavor in the process. Celery can be stored in the root cellar, and late plantings stored just before frost will be in use after New Year's. I have not experienced trouble from insects or disease, and, outside of banking, no special treatment is in order, remembering always that a lot of moisture is required.

CUCUMBERS

Cucumbers are susceptible to frost, and so should not be planted until all such danger is past. Growing them takes considerable space, but healthy plants are prolific, so it is not necessary to have many of them in a small kitchen garden. In order to save space, they can be planted at the edge of the garden adjacent to the corn, so that the vines can be trained to run in between the hills of corn. They may be planted in hills, but I prefer them in rows. The rows are prepared to receive the seed by trenching to a depth of three inches, filling the

[134]

CUCUMBERS

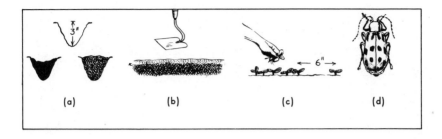

(a) Dig trench, fill with compost, mix with soil. (b) Space seeds 3″ to 4″ apart, cover with ½″ earth. Tamp down. (c) *Keep moist.* When seedlings appear, thin to 6″. (d) Spotted cucumber beetle. Handpick.

trench with compost, then mixing the earth. Seed should be scattered along this prepared row, spaced three to four inches apart, covered with one-half inch of loose earth, and then tamped down firmly. After the seedlings appear, they should be thinned to six inches.

Cucumbers require a lot of moisture so they should not be allowed to dry out. For early fruit, plant a row or two of seeds in the cold frame, transplanting them to the prepared row in the garden after all danger of frost is past. Leave a few plants in the cold frame, and although the vines will run all over the place, they will produce early, large-sized fruit.

There are several different kinds of cucumbers: small varieties for pickling and larger ones for dill pickles or table use. I plant only the larger ones, and for years have used "Straight Eight" as the most satisfactory variety. And, just for fun, "Chinese" cucumbers which grow to be two feet or so long. And their flavor is all that could be desired. As previously noted, in recent years I have planted my cucumbers along the outside rows of corn, and this system seems to work out very well indeed. Train the vines in and let them climb the corn stalks if they want to.

Cucumber beetles may cause some trouble, but can be controlled by handpicking. See under "Squashes" for explicit directions.

Cucumbers can be planted adjacent to the corn to utilize all possible space.

EGGPLANT

Eggplant is another vegetable which might well be omitted from the New England garden, but as a half dozen plants will produce sufficient fruit for a family of five, they may be included. They are a frost-tender plant and require a long season. For these reasons they are not a really satisfactory crop in this climate. The plants had best be purchased at the greenhouse and set out after all danger of frost is past. Their cultural requirements are similar to those of tomatoes. For each plant, remove a shovelful of earth, fill with compost, mix with earth, and set out

EGGPLANT

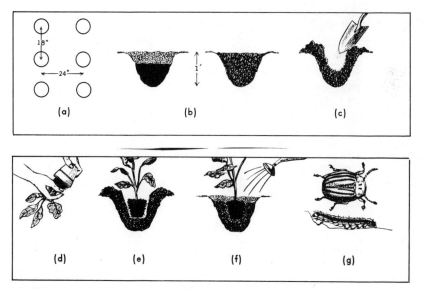

(a) Holes 1″ deep, 18″ apart, rows 2″ apart. (b) Fill holes with 8″ compost, 4″ earth and mix together. (c) Trowel out depression. (d) Remove plant from pot. (e) Set 1″ to 2″ below surface. (f) Cover with earth, water if dry. (g) Handpick potato bugs, tomato worms.

[137]

the plant. No support for the plant is necessary, and the plants should be eighteen inches apart in rows twenty-four inches apart. The enemies that I know of are potato bugs and tomato worms, both of which can be controlled by hand-picking.

ENDIVE

Endive, either the green curled or the broad-leaved, is a salad green and as such is supplemental to lettuce. Other supplemental salad vegetables are cress, witloof lettuce, chicory or French endive, and dandelions. With plenty of lettuce in the garden perhaps no other salad greens are necessary, and except for various experimental tries, I have grown none in my garden but the green curled endive. This I plant for a late summer salad and for transplanting into the cold frame for use early in the spring. It is hardy to frost, resistant to all disease that I know of, not bothered by insect pests, and can be used right from the garden after lettuce is gone. For these reasons I think it worth while to devote a couple of rows to it each year. Personal preference will decide which supplemental salad greens will be grown, if any. The culture of endive is similar to that of lettuce. The variety I use is "Green Curled Ruffec."

LETTUCE

Of all the vegetables in the home garden, perhaps the leaf crops show to the greatest advantage in taste over commercially grown vegetables, and lettuce is one of the ones where the difference is most marked. Lettuce, either looseleaf or head, grown in organically balanced soil is so superior in taste to anything to be had from the greengrocer that anyone can tell the difference. Lettuce is easily grown, affords no problems in

[138]

HEAD LETTUCE

insect or disease control, and can be had for table use from the first thing in the season right through until fall. Both the early and late plantings should be of the looseleaf variety, and the last of the late planting can be transplanted to the cold frame for use the first thing in the spring. Many people consider the looseleaf varieties superior in flavor to head lettuce, but by reason of its delicacy and difficulty in handling commercially, it is seldom found for sale in the market. Because of the advantage in taste, and since it is less trouble to grow than head lettuce, home gardeners may prefer to use two varieties of loose leaf rather than one of looseleaf and one of head, as is the rule in my garden.

Having a certain steady demand for head lettuce from my customers, I grow it each year, but do not consider it indispensable.

Being frost-resistant and liking cool, moist soil, lettuce can be the first of the small seeds to go into the garden. The first variety to be planted is the looseleaf, and it can go in as soon as the soil can be prepared. The soil should be in fine and mellow condition, and the seeds barely covered and tamped down, as described in Chapter Seven. Thinning can begin as soon as the leaves are one or two inches long, or about two weeks after they come up, for now they are big enough to be used. This thinning for table use can continue until the plants remaining are spaced about six inches apart. The head lettuce will be planted one or two weeks after the looseleaf is in and will be ready to thin by the time the thinning of the first is completed. These plants should be spaced eight inches apart in the rows, and the thinnings can be used on the table; some of them can be transplanted for heading up if there is space for them in the garden. Six or eight weeks after the first looseleaf plantings have been made, a third planting of the same variety should go in. This planting will be in use by the time the last of the head lettuce is gone and will also provide transplants for the cold frame.

Of recent years the demands of the restaurants for salads have increased and we have increased the number and varieties of our plantings. I have added "Ruby," which is in great demand for its color, though I cannot give it the highest marks for flavor. Then, too, there have been some new varieties developed, crossings between the "Boston" and "Bibb" types, I suspect. At any rate "Butter King," "Butter Crunch," "Summer Bibb," are prime among

PLAN FOR CONTINUOUS LETTUCE

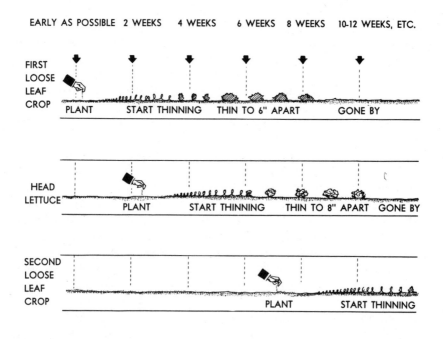

EARLY AS POSSIBLE 2 WEEKS 4 WEEKS 6 WEEKS 8 WEEKS 10-12 WEEKS, ETC.

FIRST LOOSE LEAF CROP
PLANT   START THINNING   THIN TO 6" APART   GONE BY

HEAD LETTUCE
PLANT   START THINNING   THIN TO 8" APART   GONE BY

SECOND LOOSE LEAF CROP
PLANT   START THINNING

the heading kinds, while "Salad Bowl" and "Oakleaf" supplement the supplies of "Black Seeded Simpson" in the looseleaf varieties.

Best of all though are the regular "Bibb" and any one of the several "Boston" types.

Onions

The ancient and honorable onion has a pedigree which goes back to the very beginnings of human history; it was eaten by the laborers who constructed the Pyramids of Egypt, and the soldiers of the Grecian and Roman armies were issued onions or garlic to go with their meals while on the march. All members of this quite numerous vegetable family are alleged to improve man's health, and to augment his strength and his stamina.

As a family, the onions exhibit a strong individuality; they all exude a powerful aroma, and they possess an adaptability to nearly all soils and climates. How flat and tasteless would be the sauces of great chefs without them, how innocuous our salads—and what would Thanksgiving dinner be like without a big bowl of creamed onions? From the hugest onion to the youngest shallot they are used as herbs, as medicines, as seasonings, as pickles, as salads and as cooked and fresh green vegetables. No family garden can

possibly afford to be without at least one or two members of this distinguished family.

I have grown onions from seed, from sets, and from plants; now I rely almost exclusively on the use of plants. For years I have kept a bed of "Egyptian" onions (top multipliers), have grown leeks from plants started in the cold frame, had a bunch of chives by the woodshed door, and this last year planted some shallots. Besides these there are "Welsh" onions (a bunching variety), garlic, and potato onions (ground multipliers). In the woods hereabouts, in the spring there is the wild leek, and in the fields, the wild onion.

The wild leek does not resemble the domestic plant; it has three lily-like leaves about six inches long, and from these rises a flower stalk which bears a cluster of tiny white flowers. Quite a few people here in Vermont's Green Mountains go to the woods in the early spring, while there is still snow on the ground in places, and dig up these plants for the pointed bulb which grows below. These bulbs constitute, along with dandelions, the first green stuff of the spring, and thus they have a certain virtue; personally I find them a bit strong, but they are worth trying. If you go to the spring woods to look for them, you can't miss them because of the characteristic odor. They are found in open hardwoods in the hilly and mountainous northeast.

The wild onion (which looks like an onion) is a nuisance, for it grows in the open fields and pastures and thus becomes an item in the early-spring diet of cows. The cows apparently like these shoots of green, but no one could possibly like what happens to the flavor of the milk. According to Palmer, the Indians used to harvest these plants for use throughout the year, but I am perfectly willing to let the Indians have them.

Come now, out of the woods and fields, and into the garden. Let us consider the garden varieties of onions which may be raised either from seeds, from sets, or from plants in almost all soils and climates. These onions can be harvested early for use as bunching onions (scallions). Later, after the bulbs have just begun to swell, they can be pulled to be boiled and served with drawn butter like asparagus. In the fall they're harvested as the principal crop and dried for storage after the tops have broken over. Growing onions from seed in this climate is not wholly successful, because of the shortness of the season; nevertheless, it can be done, if the soil can be prepared for them as early as the middle of April. The varieties I used to plant from seed—the "Early Grano" and "Ebenezer"— took 120 and 125 days to maturity, and I had good success with them when conditions favored early planting. Now there is a hybrid onion called "Early Harvest" which is sup-

ONIONS

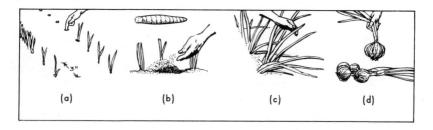

(a) Seeds: thin to 3″. Transplants: 3″ apart, (b) Control maggot with wood ash when plants 3″ high. (c) Near maturity, break over tops. (d) When tops wither, pull, dry and store.

posed to mature in 95 days from seed, but I have not tried it as yet.

The one other standard way to grow onions is by planting "sets." Sets are small bulbs which have been deliberately stunted by raising them in thickly seeded beds the year before; thus the final crop is bulbs in the second year. This is perhaps the most satisfactory method for the small garden, for the resulting bulbs, if they come from good stock, will be large and will keep well. These onions can also be culled early and used for bunching onions. In fact, this is the principal use to which they are put. The "Ebenezer" variety, either yellow or white, makes the best sets. As for bunching onions or scallions, we have a plentiful supply of these from our bed of top multiplying "Egyptian" onions.

The favored method in this climate for raising a good-sized bed of onions is to set out infant plants which have been grown in Texas and sent north by parcel post. This is the way I have grown nearly all my onions for the past ten years, and it is by far the most satisfactory way to do it. Onions from seed always require thinning and then transplanting, a tedious operation which can be avoided by the use of Texas-grown plants. Moreover the crop will be more satisfactory both as to size and succulence, for by this method "White Sweet Spanish" can be grown. These, to my mind are the best, sweet and tender and delicious, 3 to 4 inches in diameter. Further-

more this variety cannot be raised from seeds in northern climates because of the short season.

The plants come in handful bunches containing from 60 to 100 tiny plants about 5 inches long, and they can be planted just as early as the soil can be prepared. They will stand a reasonable amount of frost, and only once, when the package lay over a weekend on some post office radiator, did I find myself stuck with plants of low vitality. Ten bunches will run around 5 dollar postpaid, which is not inexpensive compared with seeds—but I've found them well worth the price.

Besides these regular garden onions, I have kept a bed of top multipliers going now for more than 20 years. Here, in my opinion, is a greatly underrated member of the family. Fantastic in appearance, ever faithful and prolific in producing a crop, requiring but little attention, giving the very first garden produce of the spring, they are surprisingly little known. I know of no one else who grows them, and this in spite of the fact that I have given away countless bulbs to people whose enthusiasm was apparently short-lived.

Since these surprising plants perpetuate themselves by clusters of bulbs which sprout from the ends of the leaves, sometimes in tiers of three, one above another, they must have a bed all to themselves. The bursting forth of these crowns of bulbs, each with its shoots of green stems, gives the plant a bizarre appearance while the stems are still erect. Gradually the stalks bend over, and the clusters of bulblets reach the ground, where they immediately take root. Because of this, the bed requires some attention to keep it free from weeds, and in the fall after the main stalks are dry and broken off, I dress the soil with wood ashes, compost and blood meal.

The only use to which we put these onions is to eat them as scallions, starting the very first thing in the spring through a period of about a month and a half—and they are the best! Actually, the bulbs, even on the mature plants, are quite small. Plants that are self-planted one summer will grow through the next, when they will produce their halo of bulblets, and then they will be pulled the following spring for table use. Our 6-foot-square bed produces more scallions than we can use. The English recommend the use of these bulblets in stews and salads, and one of these days I will bring

Leeks reaching maturity.

a mess to the house and propose that they be tried that way.

Chives and shallots need but little attention, but leeks are another story. Leeks have a long growing period, and to produce blanched stems requires that the earth be drawn up around them. But they are good, even if not blanched or trench-fed to produce exhibition bunches, and they last long into the fall, after the ground has been caked with frost and the garden sprinkled with snow. The principal requirement, as I discovered after the first attempt at growing them, is that they be started in the cold frame and transplanted into the garden. Get the seeds into the cold frame as early in the season as possible and then transplant them when they begin to crowd themselves, and before the garden gets dried out. I have discovered it pays to give leeks all the water you think they need, and then some more.

It is said that the leek was held in high esteem by the Egyptians and the Saxons, having had the status of a sacred plant, but of late years it seems to have lost a good deal of its popularity. I have found that it is well worth growing.

Of chives I have little to report, except that the bunch which has thrived alongside our woodshed door for many years has required no attention at all on my part. Its blue-green leaves with their purple blossoms are lovely to look at, and the leaves are chopped up in cream sauces and in salads add something that would be sadly missed if they were not there. I cannot account for its independence of human care and its immunity from either insects or diseases, unless it be that it seems to be in a favored spot.

As for the shallot, I planted my first bulbs last year. Like garlic, the bulb is divided into cloves, and these should be planted just below the surface. I put them in a perennial bed along with the horse-radish, and the dozen or so that I planted did very well. The bulbs, which have a more delicate flavor than onions, may be kept for a whole year, and one of my authorities reports that it is a native of Palestine.

The setting out of onion plants is similar to the planting of other transplants as far as soil and weather conditions are concerned; for the rest, the procedure is similar to that of planting small seeds, except that instead of drawing a line in the soil with the forefinger, you make a series of holes with the forefinger about three inches apart, placing a plant in each hole and firming the earth down around it.

The only trouble I have had in growing onions has resulted from the onion maggots. An onslaught of these nasty grubs can be disastrous. The plants which are not killed will be damaged by rot which starts in where the bulbs have been pierced, but since I have used wood ashes I have had no trouble. With a bucketful of fine, clean wood ashes, pass along the rows, delivering a good charge and completely covering the earth at the base of the plants. This treatment should be given as soon as the plants are two or three inches high.

PARSLEY

A little parsley goes a long way, for as the sprigs are broken off to be used as a table garnish, new shoots will come up from the center of the plant; besides that, one little bunch of parsley sprigs will do a lot of garnishing. It is worth while, however, to have part of a row in the garden and to have some in the cold frame as well. Parsley can also be taken up in the fall and potted, whence it can be used all winter long in the house.

The seeds are small and difficult to distribute sparsely, but with a little care in seeding it should not be necessary to thin them. They are very slow to germinate, the slowest of all discussed in this book, so it will be necessary to mark the row. The easiest way to accomplish this is to scatter a few radish seeds along the row; these will come up within a few days and the row will thus be marked so that the first cultivation can take place before the weeds have a chance to get a head start. Once parsley is off to a good beginning, nothing seems to disturb it. The standard variety is called "Paramount."

Parsley has not rated much in the way of space here, nor does it in garden time or space, but to me its

Parsley.

[146]

real value, in spite of the fact that I don't like the stuff personally, is away out of proportion to the space and time it demands.

If you have grown more parsley than you have use for, plants can be placed in the cold frame for use next spring. Another good trick is to pick bunches which are placed in quart mason jars, then placed in the freezer. These bunches will keep beautifully and will add garnishes for soups and salads all winter long.

PARSNIPS

The parsnip is one vegetable which we would not be without in this family, although when I was fifteen years old I would not have given a nickel for all the parsnips in Christendom. It is a root vegetable which demands rich, deep soil and which can be left in the ground all winter, with the flavor perhaps even improving as a result of exposure to freezing weather. Fall plowing rules out the feasibility of doing this, however, so they will be pulled and used in late summer and fall, or stored in the root cellar for use all through the winter until spring. They can be planted as early in the spring as the bed can be prepared, although there need be no rush to get them in the ground. The seed are of a type which nature designed to be delivered by the wind, so planting had better be reserved for a day when there is little or no wind blowing. They are difficult to place, and thinning will therefore be necessary. This should take place as soon as the plants are a couple of inches high, and the plants remaining should be placed three inches apart. This much-maligned vegetable, like all the rest, needs clean cultivation, but other than this requires no special attention, nor is it bothered by pests or disease.

As far as I am concerned this is one of the triple starred of all garden vegetables. They are wonderful on the table, they are easy to raise, and they keep beautifully far into the spring. I have always planted "Improved Hollow Crown" which is a standard variety, and a very satisfactory one.

PEAS

Peas are one of the great home garden vegetables, and in spite of the fact that they take up a lot of room and cause a lot of trouble if a good crop is to be insured, they are well worth all the space and care involved. Here again will be noticed the tremendous superiority in taste of the homegrown over the commercial article. There are two general types: the dwarf variety which needs no support, and the tall peas which need something for the vines to cling to. There is no doubt that the latter are preferable, and, in my opinion, the telephone type is the best of these. In spite of this conviction, I do not plant telephone peas, nor have I for a good many years, for this reason: they grow too tall. Telephone peas, if given ideal conditions, will grow so tall that it is impractical to try to support them. When I did grow them I used six-foot wire, which was not enough by three or four feet; as a result, the vines would fall over at the top, dropping down between the rows and forming an impenetrable jungle, so now I stick to varieties which can be grown on three-foot wire, of which I have eight rolls prepared and ready for use.

Peas are a cool weather crop and will not do well under hot and dry conditions. For this reason, peas should be the first vegetable to be planted in the spring, and as soon as the condition of the ground permits preparation.

As soon as the frost is out, and the ground is over its subsequent sogginess, set the wire, the first row eighteen inches in from the edge of the garden. After that the rows should be thirty inches apart. I like straight, evenly-spaced rows, so I use the garden line and measuring stick in setting the wire. I generally put all of my wire out at one time although the actual planting of the peas will be spread out over a period of time. After the wire is in place, trench along each side of it with the round-pointed shovel, making the trenches about five inches deep, and fill the trenches with compost or well-rotted stable manure. During all of the operation, avoid tramping on the earth any more than can be helped, planning each move so that there will be no backtracking. Now,

Little Marvel, a so-called bush pea, grows better when supported by wire.

PLANTING PEAS

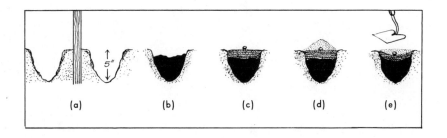

(a) Dig trenches close to wire. (b) Trench not quite filled with compost. (c) Cover with 2″ soil, drop seed. (d) Hoe 2″ light soil on top. (e) Tamp down to 1″ cover.

as the rows are to be planted hoe earth back over the compost so that it is covered two inches deep. This should be done only immediately in advance of planting; otherwise, in bright weather, the soil in which the seeds are dropped will dry out to the point that germination will be slowed down. In placing the seeds, I mark a shallow trench with the edge of my hand about two or three inches from the wire and drop the seed along this mark, spacing them anywhere between one and two inches apart. Thus there will be a double row of peas, one row on each side of the wire, spaced four to six inches apart, with the wire between them. Going between the rows of wire, two rows can be planted at a time, and then one more trip through and the seeds are covered. I prefer to cover the seed with the hoe, pulling up about an inch and a half of loose soil which is tamped down with the back of the hoe blade so that three-quarters to one inch of firm soil covers the seed.

In the first place, I had had some bad experience from seeds planted too thickly, and as a result, I swung too far in the opposite direction. With compost in the trenches and four to six inches between the rows, plants will thrive at one- to two-inch spacings, and there will be no need to thin the plants. As to the depth to cover the seeds, I had accepted as gospel much of the talk to the effect that, in order to keep the vines from drying out later in the season, it was necessary that they be planted deep. This notion I discovered to be fallacious by performing the simple experiment of pulling up a newly-sprouted pea plant. The two-inch plant had six-inch roots, and actually the fine rootlets were certainly much deeper than that, so it became obvious that deep planting could only mean slower sprouting through the surface, and could not possibly make any appreciable difference to the roots in their search for moisture.

[150]

PEA ROWS, INTERPLANTED WITH
QUICKLY MATURING CROPS

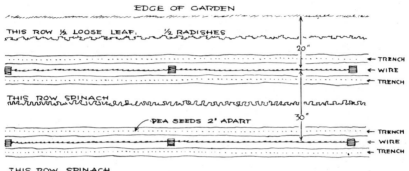

EDGE OF GARDEN

THIS ROW ½ LOOSE LEAF, ½ RADISHES

20"

← TRENCH
← WIRE
← TRENCH

THIS ROW SPINACH

PEA SEEDS 2' APART

30"

← TRENCH
← WIRE
← TRENCH

THIS ROW SPINACH

Once the peas are up, the space between the rows of seedlings and the wire should be hand weeded at least once. As soon as the vines have attached themselves to the wire, there will be no further need for weeding. If spinach is planted between the rows of wire, cultivation with the potato hook should be all that is necessary. I plant two varieties of pod peas, "Little Marvel," the earliest, and "Lincoln," which takes about a week longer to mature. In addition to this, I space the plantings over a period of about two weeks, so that harvesting is spread over a period from the last week in June until the first or second week in August.

"Little Marvel" is a so-called bush pea, and can be grown without wire, but I find that, even though it is rated as a twenty-inch vine, it will do better if supported by wire. As a matter of fact, in my experience, all peas, dwarf or not, will do better if supported.

Of recent years there has been an increasing demand for edible pod peas, called by the French "Mange Tous," sometimes known as "Chinese peas," and most often perhaps as "sugar peas." We would not think of a season now without several rows of these delicious peas, which, as the name indicates are eaten pod and all. Normally the first crop of peas of the year will be picked from these vines, and since they continue to blossom all through the season they will supply the last mess as well.

[151]

PEPPERS (SWEET)

Green peppers are nice to have in the home garden, but for the most part they are an accessory, not a main dish as far as table use is concerned, and so could be eliminated. If they are used, a dozen plants, which will not take up much space, should be sufficient for family use.

Peppers are very tender to frost, and may take as much as a hundred and fifteen days from seed to harvest; therefore it seems sensible to purchase the plants from the greenhouse and set them out in the garden after all danger of frost is past. Their planting and culture is substantially the same as for tomatoes, and as far as my experience goes, they have no enemies, except frost which, in this climate, makes the crop a hazardous one.

PLANTING PEPPERS

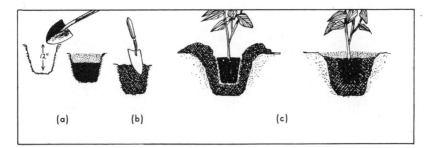

(a) Dig hole 1″ deep. Fill with 8″ compost, 4″ earth and mix. (b) Trowel out depression, remove plant from pot, set 1″ to 2″ below surface. (c) Cover with earth, water if dry.

RADISHES

Radishes are probably the easiest of all vegetables to grow, and as they will be ready to pull in about three weeks after being planted, they generally are the first fruits of the garden to come to the table. These two reasons are justification for them to find a place in every home garden, although there is no need for a wide selection of varieties or for devoting much of the garden space to their culture. If a continuous supply is desired, succession plantings can be made ten days or two weeks apart. The seed are small but round, larger than turnip seed and easier to plant. If care is taken in scattering them there will be no need for thinning, as the first roots to attain the size of a small marble can be pulled, thus making room for those left in the row. Cultural procedures are the same as for turnips, and they have the same enemies, but here again there is no danger of a major crop loss from either leaf hoppers or maggots. Along with peas, spinach, and looseleaf lettuce, they can be the earliest seeds to be planted.

On second thought, perhaps the opening sentence should be revised, for actually, along with potatoes and turnips, radishes seem to do better in soils which are not too rich in organic content, which rich condition is the aim and goal of every organic farmer. So it should be put this way, perhaps: If your soil is not in good shape, the crop of which you surely will be the proudest will be radishes. There are several standard and satisfactory varieties. The old tried and true early variety is "Early Scarlet Globe," which I have used with satisfaction for years. Of late, however, for early radishes, the only kind I plant, I have gone over to "Cherry Belles." These I think are more satisfactory in both appearance and flavor.

RUTABAGA AND TURNIP

Turnips and rutabagas are cool weather root crops, similar in growth and culture, both suitable for winter storage in the root cellar. They are easy to grow and can be planted as early in the spring as the soil can be prepared to receive them. Turnips are the first to mature, and make a good early season root crop; if they are to be used for this purpose, the seeds should be planted early. When cooked, the flesh is white and is slightly stronger than the rutabaga, which has a yellowish flesh. Rutabagas take a month longer to mature than turnips, they are much larger, and for these reasons I prefer to use them for winter storage. For home use there is no need to grow both, and unless an early summer root crop is desired, the rutabaga is my choice. Being members of the Crucifer family, which includes radishes, cabbage, etc., they are subject to the depredations of the cabbage maggot; leaf hoppers also seem to like the tender leaves early in the spring but neither seems to present a major problem.

The seed should be planted in loose rich soil in rows twelve inches apart for turnips or 14″ for rutabagas, using the method for planting as described in a previous chapter. The seeds are small but round and relatively easy to distribute. Care should be taken to scatter them as thinly as possible. After the plants are well started, they should be thinned and the thinnings can be used as green although they are not as desirable for this purpose in my opinion as are beet thinnings. They should be row-cultivated until the tops close over the rows, and the row weeding can be accomplished at the same time they are thinned. Rutabagas should be thinned to stand six inches apart in the rows, and turnips, three inches.

SALSIFY

Salsify is sometimes also known as oyster plant and in truth, when creamed or made into a cream soup, its flavor does resemble that of oysters. It is not often found for sale at the greengrocer, and that fact, coupled with its delicate flavor and its adaptability to winter storage, makes it well worth while to grow in the home garden. It is a late season root crop somewhat similar to parsnips, although it requires a slightly longer growing season than do parsnips. In spite of this fact, I have had no trouble in growing it successfully in this climate, and I would therefore assume that if it were planted early, it could be grown anywhere in the United States. It requires deeply cultivated, rich, loose earth, otherwise the roots will be twisted and malformed. The only enemy that I have discovered is the occasional presence of wire worms, which burrow into the root, but they certainly constitute no major threat to the success of the crop.

The seed are like small, round sticks tapered at the ends and about three-quarters to an inch in length, each little stick, strangely enough, containing but a single germ. Thus it is easy to place them sparingly in the row with no need for subsequent thinning. The planting and care of this vegetable is similar to that of other root crops. The roots can be left in the ground until the last thing in the fall and then dug for storage in the root cellar. As is the case with parsnips, they can be left in the ground over winter, but if the practice of fall plowing is followed, this procedure is not practicable. "Mammoth Sandwich Island" is the standard variety, and is the one which I plant.

SPINACH

To my notion, spinach is a member of the elite of the home garden group of vegetables, and it is another of those wherein the difference in taste between the homegrown and the commercially-grown is the most noticeable. Being a leaf crop, the commercial practice is to side dress with soluble nitrates so as to produce luxurious growth, which also results in giving the leaves a peculiar and inferior flavor. Some work has been done at Cornell University, I believe, in studying the effects on animals and humans which result from the consumption of vegetable foods where there is a high concentration of nitrogen present in the leaves. Unfortunately I have not been able to procure the results of these studies, and so can make no pronouncement which is not based on prejudice; but I do know that homegrown spinach which has not been fed soluble nitrates tastes better than the other kind does, and I'm sure it is more healthful as well.

Spinach is an easily grown cool weather crop, which is frost-resistant. Thus it can be planted early, and it should be, for it does not like hot weather, and germination is apt to be spotty if planted during a warm spell. If grown in rich, moist soil, the growth will be rapid and there will be no need of side dressing with nitrates; but if the soil is not rich, the growth will be slow, and the stunted plants will devote all their energy to the production of seed stalks. This production of seed stalks will also take place if the weather is too hot, so the thing to do is to get the spinach planted early if you want a good crop. Later plantings can be made for cropping during the cool weather of the fall, but for late summer and fall greens I prefer the use of Swiss chard which will be discussed later.

Spinach seed are not difficult to distribute, but care should be taken not to get them spread too thickly, otherwise time will be wasted later when the rows have to be thinned. If the seeds are properly distributed, the thinning can be for use, which takes a bit more time, but is less wasteful than if all the thinning were done at one time and when the plants are too small to be practical for use as table greens. When the plants are three inches or so high, go over the whole row pulling out

Late summer spinach thinnings.

the largest plants and those clustered together in little bunches. This thinning for use can be done a second or third time as well, if required, until the plants stand about three inches apart in the rows. With this spacing, the plants will attain their maximum leaf growth.

For the early crop, the variety I prefer to use is the "Dark Green Bloomsdale," and for the later hot weather spinach I have in the past planted "Long Standing Bloomsdale." This year I am giving trial to a new variety called "America," which is supposed to be an improvement over the "Long Standing Bloomsdale."

SQUASH

There are two general categories of squashes, the summer or bush squash and the winter or vine squash. Within these categories there are different varieties of useful and delicious squash, each one having some special claim on the grower's preference. For the small home garden, however, it is wise to omit the vine-growing winter varieties as they take up too much room, a possible exception being that, if there is a sizeable planting of sweet corn, vine-type squash may be grown in the corn patch.

Squash are tender to frost and should not be planted until all danger is past. Their culture is similar to that of cucumbers but I prefer to grow them in hills, not in rows as I have specified for the latter vegetable. Summer squash are prolific producers and two or three hills of each of two varieties should be sufficient for the family garden. These hills, which should be about three feet apart in each direction, should be prepared by removing two or three shovelfuls of earth with the round-pointed shovel and filling the hole nearly to the brim with compost. Hoe two or three inches of soil up on top of the compost and drop six seeds to the hill, covering them with three-quarters to one inch of loose soil which is tamped down to about one-half inch. When the seedlings are well started, three or possibly four plants are all that should be left in the hill.

There are a couple of beetles which love young, tender squash leaves and their depredations will probably have to be checked. These

PLANTING SUMMER SQUASH

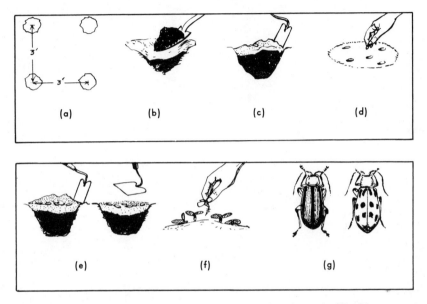

(a) Hills 3″ apart each direction. (b) Remove 2 or 3 shovels earth, fill with compost. (c) 2 or 3 inches soil top of compost. (d) 6 seeds to a hill. (e) Cover with 1″ loose soil. Tamp to ½″ cover. (f) Thin to 3 or 4 plants. (g) Striped cucumber beetle. Spotted squash beetle.

are the striped cucumber beetle and the spotted squash beetle, the latter also being called a cucumber beetle. I have never had any serious damage resulting from these pests, but always take the precaution of looking for them early in the morning before the sun warms them up. At this time, the small hard-shelled bugs are sluggish and may be picked off the plants, where they will appear on the tops of the leaves. A few moments in the morning will enable the gardener to kill most of them this way. The ones that do escape will lay yellow or orange egg clusters on the under sides of the leaves; search these out and destroy them. The few eggs that succeed in hatching will produce nasty-looking, soft-bodied larvae which feed on the leaves from the under side, and it is these bugs that do the real damage if left undisturbed. If your garden is in good organic shape, if there are plenty of birds, and if you devote a few moments each morning early in the season to handpick, you will not fail to have a good crop of squash. For control of these pests on cucumbers, the same procedure should be followed.

I generally plant three varieties of squash, all three bush varieties. Of

Gold Nugget squash.

the summer squash my preference is the "Early Prolific Straightneck," which actually is not as prolific as the "Zucchini (Elite)." Recently a new variety of winter squash has come on the market. This is sold by Harris, and by others as well, I'm sure. It is called "Gold Nugget," and a nugget it truly is. This golden squash about the size and shape of a grapefruit, is absolutely tops as far as taste goes. In addition the bush vines are compact and the prolific fruits mature early.

SWEET CORN

Sweet corn takes up a lot of space for the amount of harvest it produces; besides this it is a heavy user of plant foods; but in spite of these drawbacks, it is such a universal favorite that it must be included in the home garden. Here is another exciting instance, gastronomically speaking, of the superiority of the home-grown over the store-bought vegetable. Corn to be enjoyed at the acme of its flavor should be popped into the pot immediately after being picked. This can be accomplished by the home gardener, whereas the greengrocer's product has sometimes been kicking around as long as three or four days before it comes to the table.

Here again the climate will be the determining factor in the choice of varieties and whether or not to go in for succession plantings. In this climate, or in any section where there is a fairly short growing season, the period of harvesting can be sufficiently prolonged by the selection of, say, two varieties, each with a longer maturing period than the previous one. It is a warm weather crop and so can be grown over a long season if hot weather prevails, and in such instances successive plantings may be indicated. Here we are fortunate if we get a good harvest from the earliest varieties, so I plant my corn all at once and use two varieties, one ten days longer in maturing than the other. In an effort to get corn for use at the earliest possible moment, one may be tempted to use the very latest short-season hybrid with the result that the production of decent-sized ears and the quality of the ears are well below a reasonable standard. It is best to stick to the tried and true varieties where the number of ears per hill, the size of the ears, and the quality will all be up to standard. Here there is room for

Sweet Corn. (Eastern States sugar and gold).

[161]

wide latitude in selecting the varieties to be planted, and the catalog of one of the standard seed firms is the best guide in making the selections.

Planting can take place as soon as all danger of frost is past, and I prefer planting in hills to planting in rows. I checkmark my plot with a marker already described, the prongs of which are thirty inches apart. At each intersection I remove a shovelful of earth, fill the hole with compost or well-rotted manure, tamp it down, and with the broad-bladed hoe cover with half of the amount of soil which was removed. Compost seems to produce better growth than does manure.

Now we are ready to plant. I fill my empty righthand pants pocket with seed and start off, dropping four or five kernels in each hill without bending over. Moving along rhythmically, sometimes only three kernels fall, sometimes five; four is the ideal, but I do not stop if I miss one in either direction. Three stalks per hill is enough, and if conditions are right, nearly one hundred per cent germination will result if the best seeds are used. Later, when the corn has sprouted, the excess plants can be pulled out. Being slightly softhearted, I am inclined to leave four stalks to a hill if I find them, but five or six are too many. After the seeds are dropped, pull the balance of the loose earth over them with the hoe and tamp it down so that the seed lies under one to one-half inch of firmed-down earth. The

ideal conditions for planting are in a warm, moist soil. In cold, wet soil the seed will rot, but with good seed and proper soil conditions there will be nearly perfect germination and the shoots will break the ground in five days. Nothing will be gained by planting before the soil is just right; the seeds will take longer to germinate, and many of them will fail completely so that it may be necessary to replant.

The first cultivation should be done as soon as the shoots first mark the hills and two subsequent hoeings should do the trick, for when it is knee high it will require no further cultivation. If winter squash are desired, two seeds can be dropped in occasional hills along with the corn, these being at least five hills apart in each direction.

Here in an upland valley at an elevation of 1,400 feet—the air drainage is such that places around us but higher above the stream bed have an advantage over us amounting to a week or more on either end of the season. In this latitude (43+ degrees) the nights are seldom hot, and thus several desirable vegetables such as lima beans and melons will not reward the time and labor invested in them, even higher up on the hills. Nevertheless there are compromises to be made among the frost-tender kinds—and one of the most satisfactory of these is sweet corn.

Of all the quickly-maturing varieties of corn, most are hybrids. The

Butter and Sugar Corn.

shortest of these I could locate last season was a strain sold by Harris Seeds' Rochester, N.Y., called "Royal Crest," the package of which proclaimed that it matured in 64 days. As it turned out, this was an optimistic claim; the actual elapsed time was 71 days. My corn plot last season contained 384 hills which were planted 28 inches apart in each direction with three, occasionally four stalks per hill. Each hill had a round-pointed shovel of earth removed which was replaced with fine well-rotted, granular compost.

Two hundred thirty-two of these hills were planted to "Royal Crest," the first time I had used this variety. The balance of 152 hills was planted to "Seneca Chief," likewise being planted for the first time. Both varieties are yellow kerneled corn, and I started to pick "Seneca Chief," which purports to be a 78-day corn, I believe, on the 4th of September, just two days after the "Royal Crest" was all gone. That was 88 days after it was sown, another example of catalog optimism—at least as far as our climate is concerned.

Both varieties had been planted on the same day, the sixth of June. At this point I must admit to an error of calculation: I could perfectly well have planted a week earlier. As a matter of sound practice, it is better to plant early—for to be caught by frost on the front end is apt to be less damaging than on the tail end. Fortunately the frosts which hit us first on August 22

and subsequently were not severe, and were followed by growing weather, so the maturing of the corn was not seriously interfered with. But it could have been disastrous.

The growth of both varieties was excellent, the "Royal Crest" stalks averaging 7½ feet with reddish stems and tassels. "Seneca Chief" went about a half a foot higher with yellow stems and tassels. The quality of the "Royal Crest" ears was very fine for an early variety, the first ears running about seven inches in length and well filled to the tips. The tag end of the pickings were smaller, but all held good, sweet quality, and the crop averaged four ears per hill.

"Seneca Chief" still produces as of mid-September, and in my opinion it is the best variety that one might hope to raise in this climate; although it is not one which the gardener could safely count on. Its ears are larger than the 64-day corn, and the kernels are extra sweet and tender. In fact all who tasted this corn vowed it was the best they had ever eaten. Gardeners around here seem to prefer the mixed white-and-yellow varieties to the all-yellow ones, and of these the 78-day "Butter and Sugar" appears to be the favorite. By now, though, I have tried most of the varieties which one could hope to bring to maturity in this climate, and in my opinion "Seneca Chief" is the best of them all. Records aren't complete yet, but it is perfectly clear that its produc-

SWEET CORN, PLANTING AND CULTURE

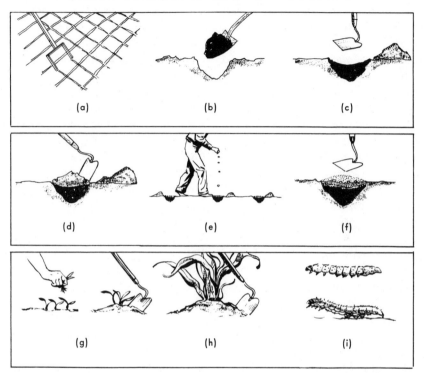

(a) Mark plot off in squares, 30″ each way. (b) Remove shovelful earth. (c) Fill with compost, tamp with hoe. (d) Cover compost with half of removed earth. (e) Drop 4 or 5 seeds. (f) Hoe over seeds with balance of earth and tamp to ½″ cover. (g) Cultivate and hill when shoots appear. Thin to 3 plants. (h) Cultivate until plants knee high. (i) Corn borer, corn ear worm.

tivity is higher, along with its superior quality. I am sure the plants will finish up at about six marketable ears per hill.

As a matter of sound practice, a crop of corn should not succeed itself the next year on the same ground. My garden is composed of four plots, each separated from the others by greensward, and the corn is rotated between them from year to year. Normally each one of these plots receives four yards of rotted barnyard manure every year. But this past fall the deep snows came before the manure could be delivered, and in the spring the ground was too soft, so I had no manure. Fortunately, I had been building a pile of organic material which I did not choose to put in my regular compost piles, and this had been going on for several years. There were cornstalks, old bailed hay, lawn clippings, shredded bark, and so forth in this pile—and at the bottom of it there was as fine compost as one could hope for. However, since there was no way to sort it out, I had it dumped on two of the quarters in its entirety, raking off the coarse material after the piles had been spread. There was enough compost in my regular piles to cover the other two sections, and so for the first time in 35 years, the garden received no fertilization except that which straight compost could give it—and I must say the results were phenomenal.

The corn quarter was one of those which received the rough compost, with a shovelful of fine compost added to each hill. (I should say now, that this last wasn't really necessary, but at planting time I could not be sure.) The corn was up in five days, and nothing interfered with its growth from that day on. The whole plot received its first hoeing by the Fourth of July, and then two more before the stalks were too high to permit further cultivation.

So, the corn grew apace, and was the marvel of the countryside—all even in height, dark-green and broad in leaf, standing in ranks so close that naught showed but a solid cube of healthy vegetation which eventually became frosted on top with the red and yellow of blossoms.

In summing up, I would say that cold-country corn is perhaps less productive than crops grown where nights are hot and humid, and that the requirement of rapid maturity excludes from the range of choice, varieties wherein the ears are larger and the kernels possibly sweeter. But in my experience it would be hard to beat "Seneca Chief." However, in this climate, this variety took 88 days from planting to picking, so it's far from being a sure thing. Thus I hedged my bet by planting more of the less acceptable 64-day corn. Nevertheless I shall be sowing "Seneca Chief" again this spring when planting time comes around, and if the weather seems propitious, it will be a week or more earlier than last year.

Swiss Chard

This member of the beet family is a well-deserved favorite for the production of succulent table greens.

Crop after crop of leaves may be cut from the plant without damage, and new leaves will continue to sprout from the center. While perhaps not regarded as highly in some quarters as spinach as a green, it will produce leaves throughout the heat of the summer at which time spinach may not do so well. It has the further advantage of being extremely resistant to cold and will be producing until the ground begins to freeze in the fall, when it can be transplanted to the cold frame where it will start out all over again early in the spring.

Its culture is much the same as for beets. The beet-like seeds are fairly large and each may include more than one germ, so care should be taken not to drop the seeds too close together. In any event, thinning will be necessary, and if the seeds have been planted sparingly, thinning for use can be practiced. As in the case of spinach, wait until the plants have achieved reasonable size, then thin as you use until the plants stand six to eight inches apart. Flea beetles or leaf hoppers will eat on the leaves to some extent, but not to the point of causing real damage. The variety I plant is "Fordbook Giant."

TOMATOES

Tomatoes are a warm weather crop, and in cold climates with short growing season, they may not be worth while bothering with. However, they are easy to grow and, as the green fruits are usable for pickling and cooking, and as they can also be house-ripened for table use, even in this climate we consider tomatoes worth growing.

The plants are very tender to frost, and if homegrown, must be started from seed in the house, with the plants being transplanted to the garden after all danger of frost is past.

I plant one packet of "Moreton Hybrid" and one packet of "Fireball" in flats in the house as I described. These are set out in the cold frame, and from thence into the proper place in the garden. The hybrids are nice big tomatoes and reasonably early, but are not as early as the much smaller "Fireballs." So the prolific smaller tomatoes serve as some sort of insurance for ripe tomatoes in the garden. In any event there will be plenty of green ones in the larger size. Tomatoes must be picked when all the signs point to frost. They can be used green, but

PLANTING TOMATOES

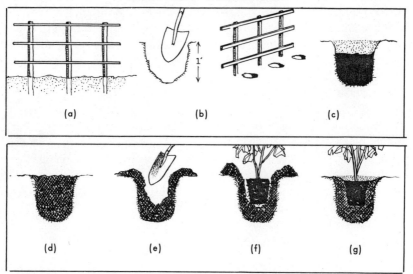

(a) Set trellis. (b) Dig holes 1' deep, 3 to a section. (c) Fill holes with 8" compost, 4" earth. (d) Mix. (e) Trowel out depression. (f) Remove plant from pot, set 1" or 2" below surface. (g) Cover with earth, water if dry.

if properly cared for they will ripen in the house or in the root cellar, and this family counts on having fresh tomatoes on until Thanksgiving day.

It will be well to keep the plants watered if there is any tendency to dry out during the transition period. This same rule applies to all transplants, and if the weather fails, the watering must be kept up until the plants are well established in their new environment.

There will be less loss from rot and insect damage to the fruits if they are kept off the ground, so some sort of support for the vines is indicated. I prefer the use of a trellis, which I have described in a previous chapter. Each six-foot section of trellis will take care of three plants. Tomatoes grown under ideal conditions are prolific producers, and a dozen plants should produce all the fruit an ordinary family will use. Leaf hoppers will feed on the leaves to a certain extent, and an occasional hornworm may be found, but the former will cause no permanent damage and the latter can be picked off as they appear.

Tomato plants need support to protect them from rot and insect damage.

PERENNIALS

Every home garden should have a plot or two set apart so that the fruits of some of the several kinds of perennials may be enjoyed. Within this category may be included strawberries, rhubarb, asparagus, horseradish, top multiplier onions, and the various herbs. This department of home gardening can be developed as room is available and time can be spared. For the purposes of this book, the growing of perennials must be considered as a sideline, and no detailed descriptions or instructions will be given. Here in Landgrove I have an asparagus bed, and a small bed for top multiplier onions, and I can testify that if time and space are available they are all very much worth while growing.

FLOWERS

A well-kept vegetable garden is indeed a thing of beauty, but the beauty may be enhanced by the inclusion of a few flowers. The culture of flowers is an entirely different art from the growing of vegetables, and I only mention here in passing that I include in my department the growing of both sweet peas and gladioli, and sometimes sunflowers, the latter a gesture in the direction of keeping the birds happy.

CHAPTER TEN

Conclusion

I THINK I WANT to end this book, which has already gone on long enough, not on a practical note as I did in a former book, but on a quizzical one. In any event, I will be brief.

In that earlier book, *How to Grow Food for Your Family*, I summed up by attempting an assessment in terms of dollars and cents, even going into discussions of canning, cooking, storing, and freezing vegetables so as to make the balance sheet as impressive as possible. Now I am convinced that as far as gardening is concerned, the practical results are the minor ones, and the major results have little if anything to do with practicality in the commonly-accepted meaning of the term. It is easy to convince a householder of the economy of changing the washer in a leaky faucet himself, rather than to call a plumber, but with the tongue of angels it might be impossible to turn him into enough of a plumber to do even so simple a job.

There are, I am sure, many people who would like to become gardeners; all that is needed is to get them started and show them the way, and I sincerely hope that this book may be of assistance in this direction. On the other hand, those who have little or no inherent love of the soil and growing things cannot be turned into gardeners no matter how strong the appeal to the practical streak may be. Thus it is that a gardening book must be addressed to those who are either potential or practicing gardeners, and as a result there are bound to be many points of disagreement which will develop, at least on the part of the practicing gardeners, as each new voice is heard on the subject. This disagreement, while often polite, is sometimes violent, and I for one am sorry that such lack of harmony exists.

To start out with, gardeners (and agriculturalists for that matter) are divided more or less into two camps: those who believe in the organic method and those who believe in the scientific method; but even within the ranks there is dissension. In *An Agricultural Testament*, Sir Albert Howard says, "Some attention has been paid to the biodynamic methods of agriculture in Holland and in Great Britain, but I remain unconvinced that the disciples of Rudolph Steiner can offer any real explanation of the natural laws or have yet provided any practical examples which demonstrate the value of their theories."

As an example of basic disagreement on the other side of the fence, let us take the case of "beets and spinach." Professor Nissley, who is extension horticulturist in vegetable growing of the College of Agriculture at Rutgers University, speaking of soil acidity in his book *Home Vegetable Gardening* says, "Beets and spinach are two good crop indicators and will readily show by their ragged or stunted growth either excessive soil acidity or alkalinity." But he does not say which shows what, so we search further. On page 107, speaking of beets, he says, "Beets, often called an indicator of soil acidity, will not tolerate a very acid soil." So that leaves us with the assumption that if the test is to be valid, spinach will not tolerate a very alkaline soil; otherwise two vegetables would not be necessary for the test.

In his section on spinach, Professor Nissley does not bring up the subject of tolerance to acidity or alkalinity, but turning to *Suburban and Farm Vegetable Gardens*, "Home and Garden Pamphlet Series Number Nine," published by the United States Department of Agriculture, when the culture of spinach is discussed we find the following statement: "Spinach will grow on almost any well-drained soil where sufficient moisture is available. It is very sensitive to acid soil. If a soil test shows the need, apply lime to the portion of the garden used for spinach, regardless of the treatment

Summer squash planted (at left) without compost and (at right) with compost.

given the rest of the area." To further complicate the issue, other authorities recommend the use of ammonium sulphate as a side dressing for spinach, and ammonium sulphate is a chemical with an acid reaction.

As I have already stated, my experience shows that both beets and spinach will do equally well on soil which is in good organic condition, and so it would seem as if the issues raised were nonexistent and that the conflicting advice given can only result in confusion. If general experience checks with what I have observed, then light is shed on another question, and that is what exactly are the functions of the chemicals which we are advised to add to the soil to feed our crops; what exactly is the need for them and what is their effect on soil structure?

Speaking of chemical fertilizers, Professor Nissley says in the same book, "Of the dozen or more chemical elements in the soil which are required for plant growth, the ground is usually well supplied with all but three which are considered the most important. These are nitrogen, phosphorus, and potassium." I would like to know why these three should be missing, and for what reason nature has failed to supply them, for it must be assumed that in nature's normal state they are present in adequate quantities; otherwise, previous to the discovery of chemical fertilizers, there would have been only undernourished vegetation on the face of the globe. On this problem I want to raise some more questions later, but for the moment let us examine the statement of fact that these elements are missing under normal conditions. With all deference to Professor Nissley, I doubt very much if such is the case. Take for example nitrogen; it is a known fact that electrical storms produce nitrogen in soluble forms which reach the earth along with the falling rain; that small amounts of soluble nitrogen are present in snow; and that the roots of all the legume family have the power of fixing atmospheric nitrogen in usable forms and storing it in their plant structures. Potassium is present in all woody growth as can easily be demonstrated by leaching out wood ashes. Potassium carbonate is the essential chemical in lye, which is made from wood ashes and from which, in the old days, all our soap was made, so the question arises, how did the woody stems acquire this potassium and why cannot it be returned to the earth as nature supplied it in the first place? As for phosphorus, who has not observed the myriad of phosphorescent insects, the glowworms, lightning bugs, etc.? Where but in nature did these insects find the phosphorus which lights their lanterns? I am convinced that the common earthworm has a high percentage of phosphorus in its makeup, and as the reason for believing this makes a strange story, I will venture to tell it.

I discovered some time ago that the large-sized, wooden-stemmed variety of matches commonly known as "kitchen matches" would blister my skin if I carried them loose in my pocket. It took a long time to discover what was making these vicious blisters on my hide, but the cause was finally and inescapably traced to the phosphorus in the matches, which would take effect through several layers of clothing. Naturally, I stopped carrying loose matches; in spite of this, however, I came up one day with a burn on my leg under my trouser pocket which had all the earmarks of being a phosphorus burn. Where had the phosphorus come from? There could be but one answer. I had been hunting that day and early in the hunt had shot a woodcock. Lacking a game pocket, I stuck it in my side trouser pocket, and there the bird had remained most of the day. The only conclusion that I could come up with was that the bird caused the burn, and the reason must have been that it contained a strong dose of phosphorus, and that probably this concentration resulted from the bird's diet, which is almost exclusively of earthworms. Of course, this is a conclusion unsupported by scientific data, but it seems to me that the incident opens up a legitimate field for agricultural research.

Now to go back to a question previously raised: Does the use of chemicals affect the structure of the soil? If it does, the fact should be well-known to agricultural scientists, and, if the structure is affected adversely, recommendations for the use of chemical fertilizers should be accompanied by a warning. In no instance, after checking over a great mass of material advocating the use of chemicals, have I seen any such warning, so the conclusion would seem to be that there is no accompanying deterioration of the soil structure. But is this the fact? Here again I raise a doubt. Actually, conversations with agricultural scientists have made it clear to me that the use of chemicals does adversely affect the structure of the soil, but this aspect of the use of chemical fertilizers is certainly not publicized. That the situation has been tacitly recognized is shown by the fact that several of the great chemical firms specializing in the manufacture of fertilizers have been working on products to be used as mechanical soil conditioners. Within the past few years, several of these concoctions have been put on the market, and their advent was hailed with a great fanfare of overoptimistic jubilation in the daily papers, in periodicals, and, above all, in advertisements.

One day a highly intelligent friend of mine burst into my study, practically beside himself with joy, having just read of this new discovery which was to revolutionize agricultural practices the world over. I was amazed at his jubilation, for it seemed to me that, even if this new product did live up to all that was claimed for it, the whole art of handling the soil for man's useful purposes would be removed one step further from the normal procedures of nature, and that in the end it might simply prove to be another case of adding two wrongs together with the hope that the answer would come out right. Subsequent use of

mechanical soil conditioners has proved, not necessarily that what the sponsors said of them was wrong, but simply that they do not work. As far as I know at the present writing, all of these materials have been withdrawn from the market, and we are left exactly where we were in the first place, and that is with a very serious problem on our hands which most of the writers on agricultural subjects choose to ignore. I hope it is not out of order for me to make a plea here that there be more frankness on this subject.

Another aspect of the use of chemicals as plant food which receives little attention, except in the field of organic gardening, is the effect their use has on the health of the plants so treated. There is a complete unwillingness on the part of the agricultural experiment stations to believe that, by the elimination of chemical fertilizers and large-scale monocultures, the problem of insect and disease control diminishes to the vanishing point. The answer seems to be, as the chap from Missouri said, "I will believe it when I see it." The results are there, however, to be seen by anyone who chooses to look at them.

Besides these questions, which are of major importance to farmers and gardeners everywhere, there are several other minor ones which I would like to hear discussed, and if the answers are not present, I would like to see the great facilities of our agri-

The Organic Method.

cultural research organizations applied to the solutions.

Exactly what are the effects of wood ashes when applied to garden soil? In all of the reference matter I have gone over in the preparation of this book, I have found practically no mention of wood ashes. If their use is helpful in making a better garden, and my experience indicates this to be the case, why has not this fact been publicized in Department of Agriculture publications? Almost any home gardener can procure limited amounts of wood ashes, and a little bit will go a long way. In almost every instance the amount and kind of chemical fertilizer to be used is mentioned, in government bulletins, but nothing is ever said about wood ashes.

There is another subject upon which I would like to see some research done, if it has not already been done, and that is, what happens to the soluble nitrates which are recommended for feeding to leafy-growth plants? Chemicals fed to plants do affect the composition of the stems and leaves of those plants to which they are fed, so the question arises to what extent are these chemicals present in the leaves of the vegetables when we eat them? And most important, if they are present in more than normal quantities, what effects do they have on the animals and humans who eat them? If there are effects, are they beneficial or otherwise? It seems to me that if the use of soluble nitrates is recommended to farmers and gardeners by public agencies, it should be done with the positive assurance that no ill effects will result.

Studies are now (1970) being carried on in this field. See Barry Commoner "Science and Survival."

That the possibility of the presence of an excess of nitrates in human diet or in the diet of milk cows is harmful is but one tiny item in the long list of questions concerning the relation between man's health and the soil his food is grown in. The answers to these questions are of vital importance to all of us, and, in view of the amounts of public funds being spent in this country by the various agricultural experimental agencies, it is high time that public opinion exerted pressure to the end that these vitally important aspects of agricultural research be given priority above all others.

In the chapter entitled "Soil Fertility and National Health" in *An*

Agricultural Testament, Sir Albert Howard barely scratches the surface of this tremendously important subject. In doing so, he brings to light much that is of great interest and significance from the studies of Mc-Carison of that magnificent race of men, the Hunzas, who live in northern India. Furthermore he quotes from the report of The Local Medical and Panel Committee of Cheshire, England which recorded their findings in "A Medical Testament" (published in the *British Medical Journal*, issue of April 15, 1939), which report has direct bearing on the relations between man's diet and health.

That the flavor of vegetables grown without the use of chemicals or commercial fertilizers is distinctly superior to that of all others can be testified to by the Ogden family and all those who use the products of our garden. That they might also be superior in health-giving qualities seems to be a possibility and one which should be investigated.

These are a few of the questions which come to mind in looking back over the material I have perused while engaged in the writing of this book. Actually, all the procedures I have written down, and all the advice I have given is the result of my own experience in working with the soil and growing things; it is not a rehash of what others have said or done. Nevertheless, down through the years I have leaned on the counsel of others, accepting from the experience of this person and rejecting the advice of that one as the results of actual garden practice dictated. Differences of personal preference are to be expected, but it seems to me that in the literature of gardening practice there are so many disagreements as to matters of fact, and of the interpretation of fact, that there should be some candid and honest reappraisals made. Having taken my stand alongside of the organic gardeners, it is obvious that I expect most of the reappraisal to come from the other side, but in admitting to this I do not exclude from my heart kindly appreciation of the devotion to the public cause on the part of all the many workers in the field of agricultural research.

Without studied intent, this book has turned out to be something else than a straightforward description of gardening practice. For better or for worse, besides being a handbook, it is a treatise wherein the author philosophizes on agricultural theory and disagrees to a considerable extent with currently accepted agricultural practice. This being the case, a word of explanation in closing is appropriate.

It seems to me that the whole pattern of our mechanized and materialistic civilization is so tightly integrated that no single aspect of it can be changed or reformed. To change any part, the whole must be changed. The overall pattern is unified and tightly knit and is the expression of our cultural values and convictions. This being the case, it is futile for me to attack any single facet and with words of fearsome prophecy predict doom and destruction. Along with automation in the factory, mechanization is inevitable on the farm. The farm worker must compete with the tenders of machines for the purchase of his necessities and his luxuries, and in order to do this he must keep in line, as nearly as he can, his production per man-hour of labor with that of the factory worker. If this be true, then, being but a single part of an unchangeable pattern, it follows that the farm factories, the large-scale food producers, are forced to seek more powerful chemicals and poisons and more highly mechanized procedures, in a never-ending search for higher production per man-hour, until Nature herself calls a halt to man's folly. Sir Albert Howard's Indore Process may be feasible in an economy wherein the standard of excellence is production per unit of land, and where there is an abundance of cheap manual labor. But in a mechanized society such as we find ourselves a part of in this day and age, a completely organic procedure is impossible, and at the very best only a modified form of it can be put into practice. In the realm of gardening, those engaged in commercial production will find themselves caught in the same dilemma. The home gardener, on the other hand, is a free agent. He is not confronted with the exigencies of price, production, and profits. He can and should treat his soil with consideration for the laws of Nature, and to do this he must turn his back on most, if not all, of the pronouncements of the latest of scientific agricultural dogma. If this be heresy, make the most of it.

Ironically enough, I close knowing that a certain amount of disagreement with the procedures that I have outlined herein will come from those on whose side I have taken my stand, the organic gardeners themselves.

Index